The Story of Flowers

First published in Great Britain in 2022
by Laurence King
an imprint of The Orion Publishing Group Ltd
Carmelite House, 50 Victoria Embankment
London EC4Y 0DZ

An Hachette UK Company

10 9 8 7 6 5 4 3 2 1

Illustrations © Charlotte Day 2022
Copyright © Mark Fletcher 2022

The moral right of Noel Kingsbury to be identified as
the author of this work has been asserted in accordance
with the Copyright, Designs and Patents Act of 1988

A CIP catalogue record for this book is available
from the British Library.

ISBN 978-0-85782-920-7

Illustrated by Charlotte Day
Designed by John Round Design
Printed in Malaysia by Vivar Printing Sdn Bhd
Origination by F1 Colour Ltd., London

Laurence King is committed to ethical and sustainable production.
We are proud participants in The Book Chain Project®
bookchainproject.com

www.laurenceking.com
www.orionbooks.co.uk

The Story of Flowers

and how they changed the way we live

Noel Kingsbury

Illustrated by Charlotte Day

Laurence King

Contents

Introduction

Flowers have had a huge part to play in a great many cultures throughout history. They appear in varying roles in many places, as materials for decoration, ritual, medicines, dyestuffs and symbols. They have attracted human interest for ever (we know that even Neanderthals buried their dead with flowers), but actively cultivating them is more recent.

This book tells the story of one hundred flowers as they came into our gardens, with an emphasis on the history of each as a cultivated plant. There is a bias towards the flowers of cool temperate climes, of Europe, North America and eastern Asia, but also some reference to warmer regions. These are the flowers we think the reader will know best, as garden plants, present in the wider landscape or common across broad regions as wild flowers. Others may be overwhelmingly more familiar as cut flowers, picked up in the corner shop or supermarket. Some may be mainly known as symbols or emblems, their stylized forms appearing on flags or badges.

What are flowers?

Nowadays we all know that flowers are about reproduction. 'Do you realize', declared the great Brazilian landscape designer Roberto Burle Marx, 'that when you give someone a bunch of flowers, you are giving them a bunch of sex organs?' And yet, in Christian Europe at any

Hemerocallis fulva,
day lily; see page 37.

6

Wisteria sinensis,
wisteria; see page 90.

rate, no one realized this until the sixteenth century; it is generally accepted that the first clear indication of gender in plants came in 1694, when the professor of natural philosophy at the University of Tübingen in Germany, Rudolph Jacob Camerarius, published a memoir concerning some experiments he had made with a variety of plants. He concluded that pollen was necessary for the production of seed, and that plants, or plant parts, could be described as male or female. The colours and scents of flowers, we now know, are all about attracting pollinators, to make them transfer male pollen to female stigma. They do this inadvertently, being essentially tricked into it by the promise of sweet nectar.

We all know about bees (and butterflies), and many of us live in countries where it is easily understood that some birds undertake pollination too. The first flowers, however, were probably pollinated by beetles, and since then many plants have evolved to aim their lures at flies, wasps, slugs, mice, bats, lizards and monkeys. It might even be argued that some plants in cultivation have evolved to be pollinated by us, in the form of breeders waving fine-point scissors and narrow brushes to transmit pollen carefully from one specifically chosen flower to another.

Flowers once cut are abstracted from the plant that grows them, to be put in a vase, strung on a thread to decorate a sacred image, or added to food; separated from the plant that bears them, they are inevitably temporary, transient, fleeting. However, cultures that learned to grow crops soon turned their hand to growing flowers, too;

Prunus x *yedoensis*
'Somei-Yoshino', Japanese
cherry; see page 86.

garden flowers will repeat (grow again) often for many years, or at least scatter their seed to ensure flowering the next year, making them a much more lasting proposition.

The beauty of flowers is to some extent compromised by their transience – we cannot enjoy them for long. Since the vast majority appear at a particular time of year, that transience is linked to an appreciation and celebration of the seasons. Many have, not surprisingly, become seasonal icons and symbols: among them tulips for spring, chrysanthemums for autumn and cyclamen for winter. Flowers themselves may come and go quickly, but in some species the specially evolved leaves around them last much longer, giving them a quality of longevity that has particularly recommended them to us. With bougainvilleas, arum lilies and poinsettias, for example, most of us are hardly aware of the flowers, only the large, colourful and durable bracts.

Flowers may have co-evolved with their pollinators, but humanity has introduced a very different set of criteria in the process of breeding. Double flowers – dysfunctional freaks of nature,

but selected and propagated by gardeners – have often been a favourite, for millennia in the case of the rose. Nature often throws up other mutations, such as white flowers, or colours that are distinct and different from the normal run of the species. These 'colour breaks' are one of the drivers of breeding for the cut-flower and nursery industry; we humans, it seems, want to have our flowers in all colours, hence 'blue' roses and 'pink' daffodils. Plants with flowers that are genetically malleable, that make themselves available in different shapes and colours, have proved particularly popular throughout history.

Our one hundred flowers appear here in a very approximate historical order, arranged alphabetically by Latin name within each period. We start with the classical world of ancient Greece and Rome, from which we have a considerable body of literature, many mentions of particular flowering plants, and even images of some preserved in Roman frescoes. We move on to the post-Roman world, including, crucially, Tang-dynasty China, one of the world's great civilizational moments. Many of the favourite flowers of this period had been in cultivation for a long time before it, but records are sketchy; in the Tang, however, we see the creation of one of the world's most influential garden traditions.

Our story then moves to the medieval period in Europe, a much maligned era when in fact real progress was made on many fronts, in gardening as much as anything else. This was also the era when the Islamic world developed a rich garden culture, which was increasingly shared with Europe. The centuries between the medieval era and the rapid changes of the nineteenth century were dominated horticulturally by the Edo period in Japan, an extraordinary time when the country turned inwards and gardening became an obsession for a relatively prosperous population. In the nineteenth century – the Age of Empire – an enormous range of plants were introduced to Europe and North America from all over the world, and gardening came of age as a mass-market hobby in the newly industrialized countries. We end in the twentieth century, when growing sophistication and consumer demand greatly increased the quantity and range of species grown.

It remains to be seen what new developments will transpire as the twenty-first century progresses. Perhaps varieties that are little thought of now will come to new prominence, not to mention the breakthroughs that will inevitably result from new developments in plant-breeding and genetic engineering. There is certainly always the appetite for flowers, both commercially and as something very close to most human hearts.

Viola odorata, sweet violet; see page 89.

Achillea millefolium (Asteraceae)

YARROW

Soldier's friend

There is a strong odour around every species of *Achillea*, which gives a clue to the plant's having a chemistry that has long made it useful. Some of this is 'allelopathic', that is to say, the plant produces chemicals that inhibit the growth of other plant species, and which also, by tasting unpleasant, help to protect it against predators. For us, the astringency and antibacterial qualities of its foliage have made it very useful as a poultice on wounds, to reduce bleeding and to limit infection; it therefore became a vital plant to have on hand for soldiers. Not for nothing did Carl Linnaeus name it for Achilles, the greatest of the ancient Greek warriors. Old names in English stress this too: soldier's woundwort, bloodwort, staunchgrass. It was used to dress battle wounds as late as the Civil War of the 1860s in the United States.

The plant's bitterness has meant that it was frequently used to flavour beer, and it was really displaced in this role only by the gradual spread of hops in the post-medieval period. Yarrow also contributed something extra to the feeling of intoxication, and there is a definite hint in many traditions that the plant could be used in a way that encouraged hallucinatory visions. Second sight or the knowledge of one's future marriage partner might be granted to those who went to sleep with a yarrow stalk poked up their nose, while for the ancient Chinese, the stalks began to be used for divination, possibly as early as the Warring States period (475–221 BC). This became formulated as the standard method of casting the I Ching, which has become the most important method of divination in Chinese culture; fifty yarrow stalks are used, preferably from as near to the home of the diviner as possible.

This particular species of *Achillea* is very widespread, across both Eurasia and North America, where it has found a medicinal use in the culture of almost all tribal people. Other species are more restricted in distribution and had more specialized uses. The damp-loving *A. ptarmica*, for example, has a milder aroma, and in sixteenth-century England it was used in salads, as a flavouring herb, and to make tea or snuff.

ORIGIN
East Asia

LONGEVITY
Long-lived perennial

SIZE
To 1 m (3 ft)

HABITAT
Woodland edge

COLOURS
Yellow, but cultivated forms can be had in every colour except blue

A single yarrow flower, one of a hundred or so to be found in a head.

The seed head of a hollyhock splits
open to produce up to ten seeds.

Alcea rosea (Malvaceae)

HOLLYHOCK

Sturdy self-seeder

Hollyhocks and the closely related mallows have been grown since time immemorial for their soft, mucilaginous leaves, which are good to eat or to feed to animals, or for use as a herbal treatment when something soothing is needed. They taste of nothing, which is why we don't eat them today. Ironically, these plants can also produce long, tough fibres, and several of their relatives have been important commercially as a source of fibre.

These are very easy decorative plants, with an amazing ability to self-seed, especially into cracks in paving. While they are only short-lived, they so often proliferate and spread, and their flowers come in such a remarkably wide range of colours, that they have become part of the personality of some places – the Île de Ré off the west coast of France being a good example.

Hollyhocks are probably of Middle Eastern origin, but they have been around for so long that it is impossible to tell. We do know, however, that the Aoi Matsuri festival, held in Kyoto every May, in which hollyhock leaves are a major element, has one of the longest runs of continual celebration in the world – since the sixth century. The leaves (but not the flowers) were used as a crest by the Tokugawa family, who established the Edo period (1603–1867) of Japanese history. The dynasty's founder, Tokugawa Ieyasu, used his hollyhock emblem instead of the imperial chrysanthemum as a way of distancing himself from the throne and the powers behind it.

A. rosea seems to have become a European garden plant during the medieval period, and it may have come to Europe via the Crusades. Several colour forms can be seen in German paintings from around this time. The early seventeenth century brought the introduction of *A. ficifolia* from Russia, bringing in genes for yellow flowers. By the eighteenth century the hollyhock had become a common cottage-garden plant across Europe. During the next century a great many varieties were selected, propagated from cuttings and given names; hollyhock rust then appeared, and in a few years these all disappeared. From then on the plant would be grown only as a seed-raised biennial.

ORIGIN
Cool temperate climates across the northern hemisphere

LONGEVITY
Short-lived perennial

SIZE
To 2 m (6 ft)

HABITAT
Open situations with disturbed ground

COLOURS
Almost all, including 'almost black', apart from bright orange and blues

Anemone coronaria (Ranunculaceae)

ANEMONE

Miracle from the Holy Land

This old genus in evolutionary terms, spread widely across the globe, encompasses some very different plants. It is one over which botanists have tended to squabble. The latest thinking suggests we may soon have more anemones, as other genera get reclassified. *Hepatica* and *Pulsatilla*, for instance, may soon lose their own identity, which will make sense to many gardeners who already know them as very similar small spring flowers.

Our example here is the blood-red spring wild flower of the Mediterranean, which would have been familiar to the ancient civilizations of the region. It became a surprise feature for some medieval Italian trading cities, too, as soil brought from the Holy Land as ships' ballast sprouted spectacular flowers, which were proclaimed to be miraculous, prompting pilgrims to carry seed all over Europe. Meanwhile the Ottomans had been growing them in gardens, and making selections and quite possibly hybridizing; by the end of the sixteenth century these had arrived in southern Europe via the same Italian trading cities. From there they spread northwards, although they did not grow well in the cool north of Europe, and gardens had to be expensively restocked every few years.

It is the hybrids of *A. coronaria* that we are familiar with as cut flowers, or as short-lived garden plants. The Dutch began to grow them commercially in the seventeenth century, but it was growers in Normandy in the eighteenth who produced the strains we grow today: 'St Brigid' and 'de Caen'.

The anemones of the European and American woods may have been in plant-collectors' gardens by the early seventeenth century, but they were always too subtle, and too slow-growing to attract much attention, although during the twentieth century they came to be increasingly appreciated by plant connoisseurs. Some southern European woodland species, however, can be treated as bulbs, and will in some circumstances eventually spread to form spectacular garden colonies in older gardens, generally around trees planted in lawns, where shade weakens the grass at the base of the trunk.

ORIGIN
Cool temperate and Mediterranean climates across the northern hemisphere, with some in the southern

LONGEVITY
Long-lived perennials

SIZE
Most under 30 cm (12 in), Far Eastern woodland types to 1.2 m (4 ft)

HABITAT
A very wide range of habitats, from woodland to open conditions

COLOURS
Reds, pinks, white, blue-purples; no real yellows, oranges or true blues

The buds of anemones are surrounded
by deeply incised bracts.

The seed head of an English marigold has rows of neatly arranged seeds.

Calendula officinalis (Asteraceae)

ENGLISH MARIGOLD

Sacred to Mary

This plant is named for the 'Kalends' – the first day of every month in Latin – because it can potentially be had in flower every month of the year. An annual weed familiar around the Mediterranean, sprouting rapidly from seed with the first rains of autumn, it was known to Roman writers mainly as a flower with which to garland statues in temples, a function that was taken up with alacrity by Hindus when it was introduced to India by Europeans at some point in the seventeenth century. (It has since largely been replaced as a temple flower by the distantly related but more productive *Tagetes* marigolds.) Along with a vast number of other flowers it was sacred to the Virgin Mary, which accounts for its name in English: 'Mary-gold'.

In fact, our *Calendula* is a hybrid of unknown and surely ancient origin, as revealed by research in 1974, between the two western Mediterranean species known to the Romans. It is a more robust plant than its parents, and, if the conditions are right in terms of temperature and moisture, it can indeed bloom every month of the year.

This was never an important medicinal herb, but it appeared in a number of magical recipes and rituals and was also popular as a culinary herb – for the vibrant colour of its flowers rather than its flavour – being used in soups, stews, dumplings and puddings. Its colour ensured that it was used time and again as a substitute for the very expensive saffron. Its availability made it useful, too, since it is an easy plant to grow. In northern Europe, seeds could be sown in spring or even autumn for a summer crop of cheerful yellow, which would carry on until the first hard frost.

The colour echoes that of the sun, and the flowers have a tendency to close up at night, an aspect that was remarked on in a couplet by the English king Charles I, who wrote while a prisoner during the Civil War in the mid-seventeenth century: 'The Marigold observes the Sun, More than my Subjects me have done.'

ORIGIN
Mediterranean region

LONGEVITY
Annual

SIZE
About 30 cm (12 in)

HABITAT
Fields, waste ground, gardens

COLOURS
Yellow and orange

Centaurea cyanus (Asteraceae)

CORNFLOWER

Royal Prussian favourite

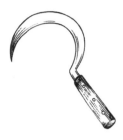

Once a common weed in grain fields, this is an arable weed that has now been almost eliminated by herbicides and the improved cleaning of seed grain. Pretty as its intensely blue flowers may have been, our ancestors must have cursed it from ancient times, since its tough stems blunt the stems of sickles – it was even known as 'hurt-sickle' in seventeenth-century England. There was also a belief that in years of bad weather, the seed of cereal crops such as wheat or rye would actually produce cornflowers rather than cereals.

Believed to have been widespread across Europe as the glaciers melted, the cornflower retreated before the advancing woodland and grassland; being an annual, it needs open ground for its seed to germinate in. However, it reconquered Europe through Neolithic farmers, who brought it back with them as they, and their practices, spread it along with their wheat, rye and barley.

The colour, pure and unusual, has long attracted people to grow it simply for pleasure, and the plant is known to have been cultivated in medieval times; by the early seventeenth century the Prince-Bishops of Eichstätt in southern Germany were growing some seven varieties, as documented in their monumental book Hortus Eystettensis.

The cornflower was never of much importance as a herb, although one sixteenth-century writer promoted it for inflammation, particularly of the eyes. The colour of the petals is fugitive as a dye and really only finds a use as a decorative element in salads. It is a good colour for symbolism, though; it became a favourite flower of the Prussian royal family during the eighteenth century, and was promoted as a buttonhole flower to celebrate the king's birthday and show support for royalism.

This is merely one of many European species of *Centaurea*, which vary greatly in their life span and habitat. The genus is named after Chiron, the centaur of classical legend, a healer who, it is said, cured his own hoof with the plant. The medieval English writer Geoffrey Chaucer records the very common knapweed, *C. nigra*, as a herb for spicing up food; it maintained a minor medicinal use, primarily for dressing wounds, until the nineteenth century.

ORIGIN
Mediterranean region, southern Europe, as an arable weed elsewhere

LONGEVITY
Annual

SIZE
About 60 cm (24 in)

HABITAT
Disturbed fertile soil in open, sunny places

COLOURS
Intense blue, with pink, dark red and white forms

The base of the flower head and, later, the seed head is distinctive and can be used to distinguish the different species.

The bud of a double chrysanthemum
is a tightly packed flower
bomb waiting to go off.

Chrysanthemum 'Heather James' (Asteraceae)

CHRYSANTHEMUM

Imperial symbol

Chrysanthemum fashions have come and gone for thousands of years. The flower has been cultivated in China for more than 3,000 years, referred to in early records usually as yellow, the colour of certain wild species. They stood out because they bloomed in autumn, after the heat of the summer, and this is one reason for their popularity ever since. It also helps to explain why they have often been linked to death; in some cultures they must never be given to people, only used for funerary decoration, and in parts of Italy a gift of the flowers is considered a curse.

By the time of the Qin dynasty (221–207 BC) in China there were almost certainly several varieties, since there was a grand market for chrysanthemum sales in the capital, while poets in succeeding dynasties wrote often in its praise. Doubles, a range of colours and multi-hued flowers appeared during the Song dynasty (AD 960–1279), with 400 varieties by 1458, when the first book on the flower was published.

In Japan, the chrysanthemum took off as a national symbol in the early thirteenth century when Emperor Go-Toba started using one as his personal symbol; other emperors followed, and late in the century it became the royal family's official symbol. During the Edo period many new varieties were produced and new growing techniques developed, often involving detailed pruning and tying to elaborate frames to shape the plants into pyramids, miniature trees or cascades, or encourage one huge, perfect flower. The latter technique was taken up widely after the plant was introduced to the West in the 1830s. Our example is one of the ball-headed varieties that hobbyist growers in Europe and North America have specialized in, and which is especially suited to the technique.

Large-scale production for cut flowers started in California in the 1880s and was later taken up by the Dutch, who today produce vast quantities for both the 'pot mum' trade and cut flowers. Growers there also lead in breeding new varieties, and the use of gamma- or X-ray radiation to induce mutations is a particularly successful technique.

ORIGIN
East Asia

LONGEVITY
Long-lived perennial

SIZE
To 1 m (3 ft)

HABITAT
Woodland edge

COLOURS
Yellow; cultivated forms can be had in every colour except blue

Crocus chrysanthus (Iridaceae)

CROCUS

Carpet of gold

A true harbinger of spring, now immensely popular as a garden plant, the crocus genus contains some 70 species. Not all are spring-flowering, however; some flower in the autumn, but do not send their leaves above ground until the spring. One of these latter is the saffron crocus, which produces the world's most expensive spice.

Saffron, *Crocus sativus*, is thought to have been first cultivated in ancient Persia, although the earliest record is that in an Assyrian document of the seventh century BC. Saffron is the stigma of the flower; it is tiny, and must be picked out by hand; some 400,000 stigmas have to be pulled from about 140,000 flowers to yield 1 kilogram (2 lb) of usable spice, representing some forty hours' work. No wonder, then, that it has always been a luxury product, hugely sought after, fetching dizzying prices and traded over vast distances. Today, in an era of chemical food colouring, exotic flavours and designer drugs, we find it hard to understand why people would have become so excited about it, but in the ancient world the intensity of its yellow, the seductiveness of its aroma and the slight buzz it is possible to get from eating a lot of it would have made it very special. Its cultivation spread beyond the Middle East, and by the late medieval period it was being grown across Europe, where place names, such as Saffron Walden in eastern England, record its spread.

Today we are more likely to spend our money on crocus bulbs for their flowers, especially if they come in rich yellows, opulent purples, subtle lavenders and rich, earthy browns, and can be scattered in huge drifts in lawn grass, to emerge magically at the end of winter, their leaves dying back tidily only a month or so after flowering. Few flowers are so easy to integrate on such a large scale into conventional gardening practices. From the seventeenth century onwards, new species arrived in Europe from the East fairly regularly, adding richer colours to the limited palette offered by European ones. By the early twentieth century the bulb industry had seen the potential of the flower, and hybridization led to bigger, stronger flowers, more vigorous plants and a wider range of colours. Crocuses are now big business.

ORIGIN
Eurasia, as far east as western China

LONGEVITY
Long-lived perennial

SIZE
Usually no more than 15 cm (6 in)

HABITAT
A very wide range, including woodland and open situations, but always with good drainage

COLOURS
Yellows, purples, off-whites, browns

The stigma (which in the
saffron crocus is the spice) and
the pollen-bearing anthers.

A carnation bud emerges from its prominent calyx.

Dianthus caryophyllus (Caryophyllaceae)

CARNATION

Flower of socialism

One of the most familiar and commercially developed florist's flowers has mysterious origins. Some people believe that it was known to the Romans, the result of several wild species of *Dianthus* having crossed and been taken into cultivation. There is certainly evidence that King Louis IX had plants brought back to France after an expeditionary force was sent to Tunis in 1270. The flower was well known in fifteenth-century Italy, and it has even been dubbed 'the flower of the Renaissance'. Further north it was rare, and thus a status symbol to be included in paintings as a sign of the wealth and discrimination of the subject; it was also increasingly used as a symbol of love.

The selection of varieties proceeded apace from the sixteenth century onwards, and being easy to propagate these were widely distributed, being found even on the windowsills of the poor. As many as 134 named varieties were listed by a writer in Berlin in 1666, classified by size. By the 1830s varieties of carnation were being bred that bloomed almost continuously. 'Sim' carnations – scentless, but hugely important for the cut-flower trade – were produced by William Sim, a Scots émigré to the United States in the early twentieth century. These were the first of many to be bred and grown in very artificial ways, to be produced in vast quantities in glasshouses and never outside, let alone in a garden. This artificiality was foreseen by the Malmaisons, first bred in early nineteenth-century France, which have a very large double centre of small petals. They are not easy to grow well, and later in the century it became the signature of a good head gardener to grow one on for four or five years to produce a plant with 50–100 flowers.

Carnations have always been associated with marriage, and in many Western countries are a typical wedding buttonhole. Another association is with motherhood, as with their use on Mother's Day in the United States. Red carnations have been used as a symbol of the socialist movement, and the flower was very popular in all colours in Communist countries. The 'Carnation Revolution' of 25 April 1974 in Portugal, however, was a more general revolt by progressive army officers against a fascist regime; carnations were placed in soldiers' gun barrels to signify their peaceful intent.

ORIGIN
Descended from species from the Mediterranean region

LONGEVITY
Short-lived perennial

SIZE
About 50 cm (20 in)

HABITAT
Wild ancestors flourish on rocks and walls

COLOURS
Dark pink through to white

Digitalis purpurea (Plantaginaceae)

FOXGLOVE

To ease the heart

One of the showiest north and central European wild flowers, with one of the largest flowers, the foxglove was always sure to attract attention. It was long appreciated as being toxic, and its reputation in many cultures was as a witches' plant, or one in some way connected with the faerie folk; this meant sometimes that it could be used protectively, but at others that it should be kept out of the home for fear of bringing in evil influences. The English name is sometimes said to derive from an old country belief that foxes wore the flowers as gloves to conceal and silence their movements.

Traditional herbalists avoided the plant, regarding it as too dangerous. However, a physician in Birmingham, in the English Midlands, discovered in the 1790s that, carefully used, extracts of the plant had beneficial medicinal properties. Soon afterwards it was found to be very helpful in regulating heartbeat; it then became a drug of some importance, and apothecaries in France even hung a painting of the flower outside their shops to advertise their skill. Digitalin drugs are those derived from the plant, and although they have been largely superseded, they were an important class of therapeutic compound for much of the twentieth century.

Such a showy and easy-to-grow plant became a popular garden species relatively early; indeed, in many places it shows up by itself. From the latter years of the nineteenth century some selection was done to produce seed strains that were predominantly white, or other colour variants. Seed companies have continued to produce novelty strains, but since the plant is a self-seeding biennial such strains are rapidly replaced by seedlings in the 'ordinary' pink.

The flower is appreciated as a wild plant perhaps more than as a garden one. Its seeds can lie dormant in the soil for decades until disturbance brings them nearer the light, which makes them germinate. Classically, there is a mass germination after trees are felled, and in those areas where woodland is maintained through coppicing, this can be a regular occurrence. The results can be spectacular, arguably one of the best wild-flower spectacles to be seen in northern Europe.

ORIGIN
Northern and central Europe

LONGEVITY
Biennial

SIZE
About 1 m (3 ft)

HABITAT
Woodland edge, disturbed woodland situations, heathland

COLOURS
Pink, with a white variant

*After flowering the ovary expands
to form a hard little seed pod,
with the style as a dry vestige.*

The seed head contains many flat
seeds, neatly stacked in five chambers.

Fritillaria imperialis (Liliaceae)

CROWN IMPERIAL

Weeping flower

There are few flowers so grand as these, although they can also be seen as absurd. Perhaps it depends on context. They are magnificent in the wild, where their dark red flowers stand out against the dry-looking grey foliage and rock of the semi-desert areas where they grow, from Kurdistan across Iran to Kashmir. As are many plants from dry habitats, they are not the most reliable garden plants, which is probably why we do not see them often.

Persia has one of the world's oldest garden cultures, and we can only assume that the crown imperial must have been one of the plants grown in the country's gardens during the many centuries of Persian civilization before the Islamic period, when it was increasingly traded into neighbouring regions, including the Ottoman Empire. From here the plant first came to Europe in 1553, having been acquired by Italian merchants. The veteran Flemish botanist Carolus Clusius was able to find it in Vienna in 1573 and take it home to be propagated and distributed around friendly princely courts. Not surprisingly, it began to appear in paintings as the ultimate floral status symbol. It also appears in a Shakespeare play, *A Winter's Tale*, published in 1611.

Early growers were fascinated by the glistening drops of nectar in the bottom of each of the flower's bells. A variety of mythical explanations were given for this, usually involving likening them to tears. One story held that this was the only flower in the Garden of Gethsemane that did not bow its head to Jesus.

As time wore on the crown imperial became more widely distributed, and by the eighteenth century the new middle classes were growing it. Opinions over it differ, and some English writers did not like the colour of the most common sort: 'boiled lobster', wrote one, and another 'ill dead orange colour'. The strong fox-like scent of the very large bulbs was another cause for complaint; this is possibly a chemical mechanism to repel the rodents that might otherwise eat the plant. The plant has certainly gone in and out of fashion, the going-outs sending the various colour forms or double varieties mentioned by some early writers to extinction.

ORIGIN
Higher-altitude areas of the Middle East, Afghanistan, Pakistan

LONGEVITY
Long-lived perennial

SIZE
About 1.2 m (4 ft)

HABITAT
Semi-desert and steppe

COLOURS
Dull orange-red, some yellow forms

Fritillaria meleagris (Liliaceae)

SNAKE'S-HEAD FRITILLARY

Enigmatic rarity

With its extraordinary chequerboard-patterned flowers and odd, dusky plum-purple-pink colouring, this is a mysterious-looking plant. With a scattered distribution across central Europe and rarity status in Britain, there are suspicions that the plant was distributed by previous generations, but by whom, when and why we do not know. Many early botanical writers fail to mention it at all, although the English playwright William Shakespeare does, and very intriguingly it is possible that the only portrait of him to have been drawn in his lifetime (on a copy of an early plant book, the famous *Herbal* of John Gerard) has him holding an outsize fritillary in his hand.

Introduced or not, the plant has flourished in Britain, in damp meadows in river floodplains, but it was picked extensively and unsustainably for London flower markets in the nineteenth century. Much of its habitat was ploughed up during the post-war period of agricultural intensification, leading to its current rare status. Its most famous location is in the water meadows of Magdalen College, Oxford, where it has become an attraction; it also features in the poems of the Victorian-era poet Matthew Arnold.

The odd colouring of the flower is on course for the genus as a whole (with one notable exception; see previous page), as many relatives are brown, grey or deep purple to the point of being black. Generally plants of the Middle East and Central Asia, where they flourish in seasonally damp habitats during the brief window between snowmelt and summer drought, they have attracted the attention of specialist growers, who describe themselves as 'frit freaks'. Another oddity is that they have enormous genomes, with some species having more DNA than any other living thing.

From the late twentieth century the species shown here began to enjoy a new life as a garden plant, because it flourishes in grass, which can be mown soon after its leaves die back. It establishes quickly and seeds relatively quickly for a bulb, allowing it to spread. A certain small percentage of the seedlings will be ivory white, making a very attractive mixture.

ORIGIN
Non-Mediterranean Europe

LONGEVITY
Long-lived perennial bulb

SIZE
30 cm (12 in)

HABITAT
Damp meadows

COLOURS
Pink-purple, white

Fritillary seed heads are the means by which many gardeners find themselves with increasing populations.

The style and anthers of
a gladiolus, normally visible
in the centre of the flower.

Gladiolus 'Atom' (Iridaceae)

GLADIOLUS

Soviet welcome

This is one of those flowers about which opinions vary, strongly. Many people love them for their cheerful colours, which are now available in an immense range. Flower-growers and florists love them because they keep well and stand straight in the vase. Others find them stiff and ugly, parading in the garden like soldiers.

In the opinion of many, 'the lilies of the field' in the Bible are in fact gladiolus, a more common and widely distributed flower in the Middle East than species of *Lilium*. A few species are scattered across the region and Europe; there are more in East Africa, and a veritable genetic explosion in South Africa.

It was the discovery of these South African species and their introduction to Europe from the late seventeenth century onwards that enabled the plants to achieve their current importance. William Herbert (see page 50) made some crosses, as an experimental amateur scientist rather than a commercial grower. In 1837 Hermann Josef Bedinghaus, a head gardener on an aristocratic estate in Belgium, made the first crosses that led to the modern flower. As more and more varieties were produced, southern Africa was scoured for new species that could be added to the gene pool, for new colours, durability and all the other features that a good garden (or florist's) plant should have. One such introduction (that of *G. primulinus* in *c.* 1890) was of a plant from the very edge of the Victoria Falls by the engineer who built a bridge across it; it brought in previously absent yellow and orange shades.

The gladiolus may be exotic and showy, but because it dies back to corms that can so easily be stored during the winter, it is an easy plant. During the mid-twentieth century it took off as a cult plant among working-class growers; specialist societies in Britain and the United States had thousands of members, who exhibited their flowers at competitive summer shows. Its short season of growth made it ideal, too, for eastern Europe and the Soviet Union, where there is in any case a strong tradition of growing cut flowers. Important guests arriving at Communist-era political events were typically met by a young woman in local costume clutching a big bunch of these flowers.

ORIGIN
Europe, eastern and southern Africa

LONGEVITY
Long-lived perennial bulb

SIZE
30–100 cm (12–40 in)

HABITAT
Open situations

COLOURS
A very wide range, but no blue

Helianthus annuus (Asteraceae)

SUNFLOWER

The cooking flower

For many of us, sunflowers are plants to grow for children. They are also important for the cut-flower industry, but are primarily an agricultural crop. Charred remnants of sunflower seeds from ancient Native American cave encampments have been pored over by archaeologists trying to work out when, where and how the plant was first domesticated. Current thinking is that this was in the region of the Ozarks (in the state of Arkansas), with hunter-gatherers collecting the protein-rich seeds to eat. Domestication reduced a large number of small heads to a smaller number of larger ones, and eventually to just one great big head. The plant spread across North and South America, and then, following the arrival of the conquistadores, by the sixteenth century it was in Spain and rapidly being taken up all over Europe – but initially only as an ornamental.

The Hortus Eystettensis of 1613, which is based on the Prince-Bishops of Eichstätt's garden, contains two magnificent pages, one with a single-headed sunflower, the other a multi-headed one. Both have yellow ray florets, but over time different colour variations would appear, and double forms. In 1762 the Berlin Academy of Sciences began to work with the government of Prussia to investigate pressing oil from the seeds and growing sunflowers on a scale that would make it a commercial possibility. The summers of northern Germany were not good enough, but the idea was taken up by the Russians, possibly at the behest of Peter the Great, a fanatical modernizer. Large-scale production eventually took off there in the early nineteenth century. Sunflowers flourished in the brief but warm summers of southern Russia and Ukraine, and provided an oil that was cheaper than other sources and with a subtle flavour that made it superb for cooking.

For many peoples, the Russians and Ukrainians in particular, the eating of sunflower seeds is a national obsession. The seeds are also familiar as an ingredient in bread and cakes, and for vegans they are an important source of protein. Over time, improved breeding has resulted in varieties that will flourish in cooler or less predictable summers, and the magnificent sight of entire fields of golden-yellow flowers, all facing the sun, has gradually spread northwards.

ORIGIN
Eastern North America

LONGEVITY
Annual

SIZE
1–3 m (3–10 ft)

HABITAT
Disturbed ground in open, sunny places

COLOURS
Yellow, with brown or dull red forms

The disc 'floret' — the flower of a sunflower — is an exercise in minimalism.

*The open, mature seed pod, with some
seeds still inside, ready to be shaken out.*

Hemerocallis fulva (Asphodelaceae)

Day lily

Oriental foodstuff

Day lilies with their open, trumpet-shaped flowers are a familiar ornamental across a surprisingly wide range of climates. They are also one of those plants of which almost every part is edible, although none of it is substantial enough to make this a real crop plant. It is as an edible plant that it has been known in the Far East for millennia, primarily for the flower buds, which are readily bought – dried – in Oriental grocery stores. Indeed, the first book about day lilies was published in 304 BC, and dealt mostly with their medicinal uses, which were various, including for treating cases of poisoning. The fact that the flowers last only a day gives them a certain value for symbolism, much exploited by Chinese poets.

While they have long been grown in Japan, little was done there to improve them as garden plants, in stark contrast to what has happened in the West, where tens of thousands of varieties have been bred. The first large-scale breeder was the American Arlow B. Stout (1876–1957), a professional geneticist and breeder, for whom day lilies were a hobby. From then the flower took off as the ideal plant for the amateur breeder, since almost every seedling from a cross looks good. A vast array of cultivars has resulted, and some 80,000 are registered with the American Daylily Society. The society's membership forms a vibrant and lively community, and, perhaps more than any other, involves its members in breeding, including laboratory-based techniques. Treatment with colchicine induces chromosome doubling, which results in plants with flowers whose colours have an extraordinary visual depth, allowing the expression of very intense colours. Those 80,000 cultivars of course need 80,000 names, and indeed a perusal of the annual list of new variety registrations reveals increasing desperation in this quarter.

The breeding of day lilies has taken the flowers in many directions, in terms of both colours, which are basically yellow through to apricot and deep red, and flower shape – narrow-petalled 'spiders', for example, doubles and wide-spreading trumpets. But there are limits. As with the daffodil, there are no true pinks, and certainly no hint of blue. Nor – and this is odd – are there any whites. And still, none has flowers that last for more than one day.

ORIGIN
Eurasia, mostly
in East Asia

LONGEVITY
Long-lived perennial

SIZE
Mostly around 1 m (3 ft)

HABITAT
Moist, well-drained
soil in grassland or at
the woodland edge

COLOURS
Yellow and orange,
with brown or dull
red and pink forms

Hyacinthus orientalis (Asparagaceae)

HYACINTH

Apollo's love

Hyacinthus was, in Greek legend, a young man whose great beauty attracted the love of Apollo, who killed him accidentally while teaching him to throw the discus. Other myths record the origin of the flower as having sprung from the blood of heroes killed in battle.

This was certainly a very early plant in cultivation, in ancient Greece and Rome, but we know next to nothing about how it was used, or the varieties available. We know it was cultivated by the Arabs in the early years of Islamic rule, whence it was taken up by the Ottoman Turks, who had a very sophisticated garden culture. It reached Europe in 1555 through the Habsburg diplomat and botanist Augier Ghislain de Busbecq. It was greatly esteemed for its scent, which is stronger than that of other late winter- or early spring-flowering bulbs and an important reason for growing it today, windowsill plants bringing the first whiff of spring into the house.

The plant grown in earlier times would have had flowers spaced out on the stem, but at some point, probably in the eighteenth century, the density of the flowers on the spike increased, ending up as the dense spike we are familiar with today. Breeding now really took off, particularly in the Dutch Provinces and German lands; by 1767 there were some 590 varieties. Such interest and diversity put the hyacinth way ahead of the tulip, the daffodil or indeed the rose. Although the passions aroused by the flower never reached the heights of the famous 'tulipmania' episode of seventeenth-century Holland, the prices some wealthy enthusiasts would pay were enormous, equivalent to many thousands of euros in today's money for a single bulb. By the nineteenth century prices had stabilized and mass production had begun; in 1815 fields around Berlin were producing 4.5 million bulbs a year, with one producer alone selling 1.5 million. In the latter part of the century production shifted to Holland, which during the next century took production away from England, too.

ORIGIN
Eastern Mediterranean region

LONGEVITY
Long-lived bulbous perennial

SIZE
About 30 cm (12 in)

HABITAT
Mostly in open, sunny places

COLOURS
Blue-purple, but with varieties in white, pale yellow, and pink to almost red shades

*The bulb is familiar to many gardeners
as an indoor plant for winter.*

The mature seed pod of an iris
contains fairly large seeds.

Iris 'Dangerous Mood' (Iridaceae)

IRIS

Symbol of France

Few flowering genera cover such a wide range of habitat as *Iris*. This, along with their unique flower shape, has given them a high profile among a wide range of cultures. The Minoans (*c.* 3000–*c.* 1450 BC) painted irises on the walls of the palace of Knossos, and the ancient Egyptians appear to have grown them as well. The ancients used the dried roots of certain species in herbalism and perfumery, and indeed orris root is still used in perfume, Arab cuisine, gin and pot pourri. These species (notably *I. germanica*) are the core ancestors of the modern bearded iris, possibly the most genetically complex of all garden plants. The incredible range of colour and pattern of modern varieties shows a legacy from a wide range of species.

The shape of the flower has led to its frequent use as a symbol. Most notable among these is the fleur-de-lis – based on the European waterside species *I. pseudacorus* – which was adopted as an emblem by the fifth-century Frankish king Clovis; it has long been a French national symbol. At the other end of Eurasia, in Japan, another species, *I. ensata*, appears on samurai family crests. Selection for the garden started in the seventeenth century, and today there are thousands of cultivars. The Higo (Kikuchi) clan on the island of Kyushu grew them, and had an iris society with strict rules – in particular that members were forbidden from giving the plants to outsiders.

Systematic breeding of the bearded iris did not begin until the early nineteenth century, in France and Germany, and later in Britain and the United States. The nineteenth-century Frenchman Jean-Nicolas Lémon was a particularly successful breeder who had the bright idea of naming his irises to capture the imagination of the gardening public: after classical gods and heroes, popular novelists, the characters in their books, or opera singers.

There are several other classes of iris, some better known as cut flowers, such as the Dutch iris group and the so-called Siberian irises, derived from the Eurasian *I. sibirica*. These long-lived, easy garden plants came into vogue in the 1920s and have been garden stalwarts ever since. North America has its own irises, and the Pacific Coast and Louisiana groups have their own aficionados, societies and expanding lists of varieties.

ORIGIN
Temperate and semi-desert climates in Eurasia; a more limited range in North America

LONGEVITY
Long-lived perennial

SIZE
15 cm–1.5 m (6 in–5 ft)

HABITAT
A very wide range, from waterside marginals to semi-desert and steppe

COLOURS
Almost the entire spectrum; even wild plants cover everything from white to black

Jasminum sambac (Oleaceae)

JASMINE

To perfume tea

'Jasmine', a bit like 'lily', is a name applied to many flowers. Ours is the 'Arabian jasmine', one of the 200-odd species of *Jasminum*, the true jasmines. There are many other climbing plants with sweetly scented white flowers that have evolved to be pollinated by moths, and many of those also get to be called jasmines.

J. *sambac* is probably of Indian origin, since there are many varieties there, mostly not known in the West. They are widely used for Hindu *darshan* (worship rituals), such as the making of garlands for statues of the gods, or in wedding celebrations. The production of jasmine, from this and *J. grandiflorum*, has become an important economic activity for small farmers in the southwest Indian state of Karnataka, mostly to be distilled for perfume. Chanel No. 5, created in the 1920s in collaboration with the French fashion designer Coco Chanel, relies heavily on Indian jasmine. Some 8,000 flowers are needed to yield just 1 gram (0.03 oz) of oil.

Jasmines are esteemed in Middle Eastern cultures for their scent, and there is some evidence that they were grown by the ancient Egyptians. They have long been used in the Far East, too, and perhaps the most familiar application is the Chinese custom of using them to flavour tea. These two regions were of course linked via the Silk Route, and it is thought that jasmines and jasmine-scented products have been traded along it since time immemorial.

The hardier *J. officinale* probably originated somewhere in a belt from Iran to northern India, but it has been in cultivation so long that the date and location of its introduction are lost. It possibly arrived in Europe at the time of the medieval Crusades. Very widely cultivated, it is one of those plants that is impossible to improve, so there are few cultivars. *J. polyanthum* is also very familiar, being the one that is mass-produced for selling trained over a circle of wire as a winter pot plant. Not quite hardy enough for most garden situations in northern Europe, it will, in sheltered places or conservatories, grow luxuriant and fill the air with a deliciously heady scent. It was originally found in Yunnan, southern China, and introduced in 1931 by Lawrence Johnston, creator of the famous garden Hidcote in the English Midlands.

ORIGIN
Eurasia, mostly in more southerly, warmer regions

LONGEVITY
Long-lived climbers and shrubs

SIZE
Some species can climb to 10 m (33 ft)

HABITAT
Woodland edge

COLOURS
Most are white; some yellow and pink species

Two buds and a cross section of a
jasmine flower. Note the long calyx tips.

A single flower, one of many arranged vertically in spikes.

Lavandula x *intermedia* (Lamiaceae)

LAVENDER

Grown for the laundry

A pungent but attractive odour marks lavender out. Of all the oil-producing herbs, this is the one that people (Europeans at least) have rated most highly over the centuries. We think the ancient Romans grew it, although there is little evidence, since it is quite likely that the lavender they are known to have used in the laundry was simply gathered on the hills. The name comes from the Latin *lavare* (to wash); clothes scented with it smell clean (even if they are perhaps not!), and bunches of dried stems help to keep insects away from clothes in storage. It has also long been rated medicinally, although with little evidence for any efficacy beyond the psychological.

Central and northern Europeans were the pioneers in growing lavender, for they had none growing wild. Medieval monasteries were the first cultivators, but for the drying of its foliage and the distillation of its oil, rather than as an ornamental. The twelfth-century German abbess Hildegard von Bingen mentions it first, and it is known to have been widespread by the sixteenth century. As European societies became wealthier, the demand for lavender grew, and alongside lavender farms – usually on dry, sandy soils – it began to appear more frequently in gardens. Different species were introduced, although their hardiness varied. A breakthrough occurred at the beginning of the nineteenth century when the hybrid *L.* x *intermedia* first appeared, we do not know exactly where or how. Despite not being as hardy as *L. angustifolia*, which still dominates northern European gardens, it is more vigorous and yields more oil.

Lavender has a long and strong association with love, and has been promoted as an aphrodisiac, as well as a token of love. In Elizabethan England a small bunch of lavender was used much as we would long-stemmed red roses today, to express romantic intent.

The use of lavender as a garden plant took off in the eighteenth century, when it began to be used to edge borders. Small forms work best for this, and from then on dwarf varieties began to be selected, among which 'Munstead' is the best known, a selection made by the celebrated plantswoman Gertrude Jekyll (1843–1932).

ORIGIN
Mediterranean region, into southeastern Europe

LONGEVITY
Subshrubs surviving for about twenty years

SIZE
To 80 cm (32 in)

HABITAT
Open, sunny places, including on very poor soils

COLOURS
Violets, purples, some near blues

Leucanthemum vulgare (Asteraceae)

OX-EYE DAISY

Queen Margaret's device

We all love daisies. Their sunny rays seem to invoke a positive feeling in everybody. This shape, and the form of yellow centre and white outer segments, is repeated across a very wide range of genera in the wider daisy family – the botanists' Asteraceae. Our example is a very common species across Eurasia and has been introduced to North America, too.

Daisies, however, despite their ubiquity and diversity, are oddly of little interest historically. Our ancestors must have been familiar with many, but few are remarked on or recorded in art, and little use was found for them medicinally, although they had some use in treating wounds. One early reference concerns Queen Margaret of Anjou, wife of the English king Henry VI in the fifteenth century, who had them as her heraldic device, embroidered on her robes and those of her attendants. The flower was often known as the 'marguerite', which is the French word for it but in English is applied only to the shrubby Canary Islands species in the genus *Argyranthemum*.

The far smaller, but perhaps even more familiar daisy, *Bellis perennis*, is known to us as a lawn weed, but also as a garden flower from the early fifteenth century, with varieties in different colours; much of the pre-industrial era interest in growing these daisies was in the freak forms that occasionally occurred and were propagated as curiosities, such as doubles or forms that send out miniature daisy heads around the outside of the main head. As a garden plant it is one of those, and they are few, that appears to have been far more common in the past than it is at present. Its small stature made it ideal for the tiny gardens that were normal for most townsfolk in the past. It was ideal for edging or for containers on windowsills.

The far larger plants known as Shasta daisies (*Leucanthemum* x *superbum*) are closely related but not descended from any of these. Bred by the prolific horticulturalist Luther Burbank (1849–1926) in California, they are of mysterious origin but have proved immensely successful, their size making them the right scale for modern gardens and their persistence and robust nature ideal for the contemporary desire for minimal maintenance.

ORIGIN
Cool temperate climates across Eurasia, introduced to North America. Very closely related plants native to North America

LONGEVITY
Short-lived perennials

SIZE
Height to 60 cm (24 in)

HABITAT
A very wide range of habitats, mostly in open, sunny places

COLOURS
Off-white, but with occasional pink or crimson forms

A single disc floret, with
flour miniature petals.

The stamens and style, the male
and female organs respectively,
at the core of the lily flower.

Lilium candidum (Liliaceae)

LILY

Symbol of purity

'Lily' is one of the most confusing flower names, since a vast number of unrelated plants are landed with it. There are about 100 species of true *Lilium*, although the boundaries are much disputed by botanists. The one shown here, *L. candidum*, is the Madonna lily of Christianity, although it is known from the frescoes of the Minoan civilization, some 1,700 years BC. Its origins are obscure, since it was widely traded by the ancient peoples of the Mediterranean and Middle Eastern regions. The purity of the white of its flowers made it a great favourite for religious symbolism, and the association with the Virgin Mary became particularly strong.

The other lily of pre-modern Europe, *L. martagon*, is a very different plant, its dark pink, spotted petals reflexing in a way that flowers very rarely do. Dubbed the 'Turk's cap' lily after the turbans worn by the Ottomans, it was extensively cultivated in the gardens of the wealthy after its introduction in the late sixteenth century.

Along with the extensive association with the sacred, the lily was put to mundane use by our ancestors, too, as a cure for corns and boils. In the Far East, several species are eaten, and in Japan whole fields of lilies can be seen growing as a food crop, notably *L. auratum*, one of the most spectacular of all, which caused a sensation when it was brought to Europe in the mid-nineteenth century. The Japanese were well set up to feed European and American gardening consumers, and in the late nineteenth century a thriving nursery trade got going around Yokohama to feed Western markets. During the previous two centuries Japanese growers had done some hybridizing, laying the foundations for the very extensive breeding work that was carried out in the next century, thanks to Jan de Graaff (1903–1989) in Oregon, United States.

Not as refined as 'the Madonna', *L. regale* is white, heavily scented and a better garden plant. The plant-hunter Ernest Henry Wilson found vast numbers flowering in a remote valley in the border region between China and Tibet. Plant-hunters brought back many more species, mostly from southern China, but many also from the western United States. Some have proved difficult in cultivation, but others have shown themselves to be more tractable.

ORIGIN
Cool temperate climates across the northern hemisphere

LONGEVITY
Perennials, but not necessarily long-lived

SIZE
30 cm–2 m (12 in–6 ft)

HABITAT
A very wide range of habitats, mostly in light shade, often in mountainous regions

COLOURS
Everything but blue

Narcissus 'Tête-à-Tête' (Amaryllidaceae)

DAFFODIL

Spring's messenger

This is the European spring flower par excellence. Our example is one of the most successful plants of all time, bred in the 1940s by Alec Gray, a fruit-grower and amateur plant-breeder. By 2006 it made up 34 per cent of Dutch bulb production. Gray's innovation was to brave post-Civil War Spain to collect naturally small wild varieties to use in his breeding work.

Daffodils had been at the cutting edge of plant-breeding work before. At the turn of the nineteenth century William Herbert, a lifelong enthusiastic plant-breeder, made a study of daffodils, showing through experimental breeding that they hybridized naturally. This contributed to his developing a version of the theory of evolution, decades before Charles Darwin. Another country cleric, George Engleheart, later in the century, played a crucial role in the development of the modern daffodil; his 'Will Scarlett', with its dramatic orange cup, was quite unlike anything else that had been seen, and led to a whole new vein of breeding. Daffodil-growing took a leap forwards in the late nineteenth century, when two key British gardeners, William Robinson and Gertrude Jekyll, showed how easy it was to plant them in rough grass and watch them come up year after year. This helped to stimulate major commercial development in the century that followed.

Daffodils have long been one of our favourite flowers. The white, heavily scented *N. tazetta* has been found in ancient Egyptian tombs and was mentioned by classical writers: Homer, Virgil and Ovid. The Silk Road took it to China, where it has long been used in the Spring Festival. Pockets of it naturalized all along the route.

The botanical name commemorates the Greek legend of Narcissus, who fell in love with his own reflection; this is also possibly a reference to the plant's supposed (although not well documented) narcotic properties. The scent of some species is indeed so strong that people can be overcome by headaches. The range of species is wide, and includes a number of flower shapes, although all have the distinctive trumpet-like corona, which early twenty-first-century research indicates is unique to *Narcissus*.

ORIGIN
Mediterranean region, with most species diversity in Iberia and the Maghreb

LONGEVITY
Long-lived perennial

SIZE
10–40 cm (4–16 in)

HABITAT
A very wide range, mostly in open, sunny places, often in mountains

COLOURS
Yellow; some white species and varieties

*The ovary and style of a daffodil,
at the core of every flower.*

*The carpellary receptacle of
a lotus is a unique structure
bearing several female stigmas.*

Nelumbo nucifera (Nelumbonaceae)

LOTUS

A Hindu throne

The lotus is one of the most extraordinary flowers, a big, complex bloom that emerges from the mud of shallow pools or marshes alongside dinner-plate-sized leaves. Its roots are edible, too, while its seeds can remain viable for centuries.

The flower's beauty has led to it holding major symbolic significance for Eastern religious and philosophical traditions, dating back to the Indian Vedic period (to 1,500 years ago). Hindu deities including Brahma, Saraswati and Lakshmi are often portrayed sitting on a stylized lotus throne, while the flower appears in the iconography of many others. Every aspect of the flower and plant has been exploited for its symbolism, such as the unfolding petals representing the expansion of the soul, while the drops of nectar on the central style of the flower are likened to jewels. The style is in fact unlike that of any other flower, which is one of several pointers to the lotus's truly ancient origins, for this is one of the earliest evolved of all flowering plants, and fossils from the Cretaceous period (145 million – 66 million years ago) have been found in Japan.

For Buddhists, the lotus represents purity of body and mind, floating above the mud of desire and attachment. People of the eclectic Bahá'í faith commemorate the flower in the dramatic shape of their headquarters and temple complex in New Delhi. The flower appears time and again in a range of secular national and political contexts, too, often as an element in flags, insignia and symbols.

All parts of the plant are edible, which adds to their atmosphere of benevolence. Lotus root has long been eaten in Asia, appreciated for its crunchy texture and sophisticated flavour. The seeds are often made into a variety of products, including Chinese festive moon cakes.

A recent use of the plant is for water filtration. Research shows that it is very good at extracting toxic heavy metals from water – so-called phytoremediation – as well as helping to reduce nutrient pollution by providing a home for beneficial bacteria on its roots.

Gardens in the Far East have always made use of the lotus, but the plant is increasingly popular in Europe and North America, since it can be relatively easily protected from winter cold; many different varieties have been selected, too, including dwarf ones.

ORIGIN
Subtropical and tropical climates in the northern hemisphere, from Egypt to Japan. There is a related species in the southern United States

LONGEVITY
Long-lived perennial

SIZE
Up to 1 m (3 ft) above water level

HABITAT
Shallow water bodies

COLOURS
Many shades of pink; cream to pure white

Nerium oleander (Apocynaceae)

OLEANDER

Beautiful but deadly

Growing wild in stony riverbeds but actually remarkably adaptable, oleander is a distinctive Mediterranean shrub. Depicted in frescoes on the walls of the ruined city of Pompeii, it was known to the ancients not just as a flowering plant but also as potentially very poisonous. It could be used in small quantities as medicine, and in larger for poisoning vermin or even wolves. Its bitter flavour has meant that human poisonings are rare but not unknown; using the long, straight twigs for barbecuing meat has been a route to death more than once. As with any very toxic plant, however, it can be hard to separate myth from reality. It is difficult, for instance, to believe that soldiers in the Napoleonic Wars actually died from sleeping on pallets made of oleander branches. But it has been suggested that the Oracle of Delphi's hallucinatory visions were brought about by ingesting calculated doses of oleander extract.

The botanical name is thought to derive from the leaves' similarity to those of the olive (*Olea*). Oleander reached England and the German lands by the sixteenth century as one of the exotic evergreens that only the very wealthy could afford the space to keep indoors over winter. As time went on the number of colour forms increased, and some double varieties appeared. In the nineteenth century plants were introduced from India, many of them yellow-flowered and more strongly scented. Being easy to propagate, they became popular during the century across central Europe, since even humble householders could find somewhere to overwinter them as pot plants, putting them outside in a sunny spot in spring, to provide several months of flower. By the mid-nineteenth century some fifty varieties were available.

An immensely important plant for gardens and landscapes in warm climates, the uses of oleander run from the mundane (high central-reservation planting) to the high-end (new dwarf varieties as balcony or terrace plants). They are among the most popular urban-landscape plants in warm-climate cities, and there are so many in Galveston, Texas, that the city has an annual oleander festival.

ORIGIN
Mediterranean region, across to northern India and southern China

LONGEVITY
Long-lived shrub

SIZE
To 6 m (20 ft)

HABITAT
Naturally in riverbeds and floodplains

COLOURS
White, pinks, reds, some yellows

The narrow seed pod, split open to reveal the seeds with the silky fibres that allow them to blow away on the wind.

The fruit, containing seeds that float
for a number of days before sinking.

Nymphaea 'Attraction' (Nymphaeaceae)

WATER LILY

Painter's obsession

The endlessly popular water lily dominates planting design in open water. There is very little that can rival it for what it does. The Latin name, *Nymphaea*, is derived from the mythical Greek beings who were often associated with water. Water lilies are among the most primitive of all flowering plants, since fossils have been found in Jurassic rocks (201 million – 145 million years ago) and many species are thought to have changed little since. During warm periods the genus became very widespread, leading to some populations becoming 'stranded' and then going off on their own evolutionary path; a 'recent' (in geological terms) example is offered by those in lakes around hot springs in Hungary, which survived the Ice Age from a previous interglacial period.

Several species are found in Egypt, where they were symbolic for the Nile civilization as a metaphor for creation, emerging from mud into sunlight. They were depicted in frescoes on tomb walls, and remains have been found in pharaonic tombs; it is believed that they featured in decorations used during religious ceremonies. A water-lily motif also appears on temple columns. Given the similarity of both flower and plant to the lotus (see page 53) – also very primitive, but not very closely related – such historic representations can be confusing to interpret.

Water lilies in gardens are often hybrids, many bred in the nineteenth century by the Frenchman Joseph Bory Latour-Marliac. He produced about 100 varieties, initially as a hobby, but later realizing their commercial importance and subsequently naming only sterile selections, so that other growers could not obtain them from seed. He produced varieties that grew at various depths, including dwarfs. Tropical water lilies have also been extensively and gloriously hybridized.

The flower is particularly associated with the prolific French Impressionist painter Claude Monet, who in 1883 bought a house at Giverny outside Paris and dedicated himself increasingly to gardening, including making extensive ponds in which he grew wild and hybrid water lilies. Many among his final series of paintings, which record his progressive loss of sight, feature his ponds and water lilies on vast, highly atmospheric canvases.

ORIGIN
Global, in tropical to cool temperate climates

LONGEVITY
Long-lived perennial

SIZE
The flower will grow to the depth of the water, up to 1 m (3 ft) deep; dwarf hybrids need just 30 cm (1 ft)

HABITAT
Still fresh water

COLOURS
White, pinks, blue-violets, pale yellows

Paeonia 'Sarah Bernhardt' (Paeoniaceae)

PEONY

Symbol of wealth

The large, lush, often richly coloured flowers of peonies have long attracted our attention. It is likely that our ancestors valued them initially as medicine, for the first time they are mentioned in myth and documents it is for their healing properties. The Roman poet Virgil, for example, records that the goddess Artemis was brought back to life by peony after being killed by her father's horse.

The Chinese were first to value the peony for its evanescent beauty, as early as the Northern and Southern dynasties (AD 420–589), and since then it has played a central role in their iconography and garden culture. Peonies symbolized wealth, so to have a garden of them was a fine example of 'conspicuous consumption'. The flower appears more frequently than perhaps any other on Chinese ceramics, although often so stylized as to be unrecognizable. These ceramics, and their peonies, were widely imitated by Western producers from the eighteenth century on, making the peony likely to be the most common 'flower of decoration' after the rose.

The peonies favoured in the Orient were what we now call 'tree peonies', i.e. they are shrubby. European species are herbaceous, disappearing over the winter. Some began to be collected and grown in the medieval period, although being slow-growing and not easy to propagate they remained the province of herbalists and the wealthy. Somewhat ironically, it was a herbaceous species from China – one that had been of little interest among the Chinese, *P. lactiflora* – that really took off in the West. After its introduction in the late eighteenth century, nurseries in Belgium, England and France began to select varieties and make crosses with other species. It was the Frenchman Victor Lemoine, the most prolific plant-breeder of all time, who in the late nineteenth century produced a range of hybrids that took Europe's gardeners by storm, including one named after one of the greatest celebrities of the age: the actress Sarah Bernhardt. Further adventurous breeding was carried out in the next century by the American A. P. Saunders, who was able to introduce reds and yellows into the gene pool. Breeding took another leap forward with the opening up of China in the 1980s and greater access to the immense array of varieties there.

ORIGIN
Temperate climates
across Eurasia

LONGEVITY
Long-lived perennial

SIZE
30 cm – 2 m (1–6 ft)

HABITAT
A wide range, including
woodland edge,
grassland, steppe,
mountainside

COLOURS
Primarily pinks;
also white, soft
yellow and scarlet

The immature fruit of a peony, with
its distinctive horns. It will split
longitudinally to release the seeds.

The distinctive ovary matures into the pepperpot-like seed capsule.

Papaver rhoeas (Papaveraceae)

POPPY

Symbol of remembrance

Poppies are bold flowers. Our example, the field poppy, traditionally spattered its scarlet flowers across cornfields. It, like all the annual members of this genus, thrives on disturbance, its seeds lying dormant for decades (sometimes even centuries) until the right conditions for germination arise. It flowered on a huge scale along the Franco-German border after the destruction of World War I, and in Britain became a symbol of this and all wars; poppies are still sold every autumn to raise money for military veterans. As a garden plant it is known through the breeding efforts of the Reverend William Wilks of Shirley, southeast of London, who in the 1880s gathered seed from any plants that did not have red flowers, developing a much-loved seed strain.

It is the larger opium poppy (*P. somniferum*) that has been the most important plant for humanity. The oldest representations of this plant date from the Sumerian civilization, about 4000 BC. They, and other ancient civilizations – and indeed almost certainly hunter-gatherers before them – had realized the pain-relieving properties of the opium that is so easily extracted from the seed heads. The plant continues to be grown on a large scale today, for the production of both therapeutic drugs and illegal ones; morphine, codeine and heroin. It has also been popular as a garden plant, with a variety of colour forms and some spectacular doubles.

The Iceland poppy, *P. nudicaule*, is a wild flower of the Arctic tundra, although ironically not of Iceland itself. It arrived in cultivation in 1730, initially to England, and eventually became successful as a garden plant and even for the floristry industry, since the flowers last for several days in the vase. Now, after much commercial development, it is available in a gorgeous array of warm colours, as well as the predominant white.

Some poppies are soundly perennial. Two of those, both from Turkey eastwards – *P. bracteatum* and *P. orientale* – were introduced to European cultivation in the early eighteenth century. From the early twentieth century onwards, they were crossed to produce a race of plants that are much loved for the bold flowers they produce in early summer, year in, year out.

ORIGIN
Temperate climates across the northern hemisphere

LONGEVITY
P. rhoeas is an annual; others are biennials or perennials of varying life spans

SIZE
20–80 cm (8–32 in)

HABITAT
Open, sunny places; shorter-lived forms on disturbed ground

COLOURS
Reds, pinks, white; some species yellow to orange

Ranunculus asiaticus 'Picotee Pink' (Ranunculaceae)

RANUNCULUS

Discovery, loss, rediscovery

Small pots of flowering ranunculus are widely available in winter and spring. There is a whiff of the artificial about them: the almost oversized flowers and vibrant colours, and the fact that they have such a brief season – planted out in the garden afterwards, they never seem to reappear. They did in fact disappear from commercial horticulture for some time, but the twenty-first century has brought a revival. Their roots go back a long way.

As plants of the Middle East, enjoying a brief spring flowering before retreating to small tubers underground, they were first grown in the Islamic world. We know little of when, but it is possibly a very long time ago indeed. They came to Europe via the Ottomans in the sixteenth century, being in fact an item of trade, although the roots that arrived in Europe were often too dry and shrivelled to flourish, and those that did rarely matched the descriptions the merchants had been given. Once growers, mostly in what is now Flanders and Holland, began to raise their own from seed, they took off commercially.

Ranunculus joined the select band that became known as 'florist's flowers', the florists being amateur growers – usually skilled artisans – who grew certain flowers as a hobby organized around competitive exhibitions. This encouraged innovation in those plants such as ranunculus that often produced 'colour breaks' (new colours) but also a kind of conformity, as classes were established for different types, and plants had to match the definitions for each. The movement developed in the Low Countries in the seventeenth century and was taken to England by refugee weavers during periods of religious conflict in the region. By the late eighteenth century 800 varieties were listed in England. Those with striped or streaked flowers were rated most highly.

With so many new plants being introduced in the nineteenth century, the ranunculus fell from grace, and, not being hardy, it could not survive in northern European gardens. It was only at the very end of the twentieth century that Dutch bulb-growers again realized its potential, albeit as a 'here today, gone tomorrow' plant.

ORIGIN
Middle East

LONGEVITY
Short-lived perennial

SIZE
To 40 cm (16 in)

HABITAT
Fields and open landscapes

COLOURS
Originally yellows and oranges, but in cultivation everything from white to very dark brown-red; no blues

*The tuberous roots of this particular
species allow it to be dormant
during hot, dry summers.*

The seed capsules, which contain many tiny seeds, are prominent in the summer after flowering.

Rhododendron luteum (Ericaceae)

YELLOW AZALEA

Sweet poison

The dividing line between 'rhododendrons' and 'azaleas' is a blurred one, and not one that makes much sense to the botanist. In common parlance deciduous rhododendrons are always dubbed 'azaleas' and much appreciated for their colourful flowers and strong, sweet scent. That scent, however, masks something sinister.

The fourth-century BC Greek soldier and writer Xenophon was one of many who reported poisonings caused by eating honey made from yellow azalea nectar. His soldiers suffered while trying to invade part of what is now Turkey. None of his men died, and indeed human fatalities are rare, although domesticated honey bees can sometimes be killed by it. The active ingredients are grayanotoxins, which interfere with nerve messaging in the brain. Hallucinations are frequent, and indeed azalea honey has been – and in some places still is – sold as a recreational drug, and as a sexual stimulant (for men only!). The toxins are produced by many plants in the heather family (Ericaceae), to which the yellow azalea belongs.

The plant was brought into cultivation in 1701 by the French botanical traveller Joseph Pitton de Tournefort. A century later a baker, one P. Mortier, in the Belgian city of Ghent crossed it with more recently introduced North American species to produce highly scented plants with flowers in a wider range of yellows and pink-reds. Others followed, and the plants became a local speciality, flourishing on the poor, acidic soils of northern Flanders. English nurserymen produced further hybrids from these, then in the 1830s Dutch nurseries worked with the Japanese species *R. molle* to produce the 'Mollis' hybrids. Later work in England resulted in the 'Exbury' strain, noted for their rich colouring, in which much of the gene pool came from the western American *R. occidentale*.

Much of the most effective azalea-breeding happened in a region of poor, rather acidic soil in Surrey, southeastern England. Traditionally something of a wasteland, with only rough grazing, and famous for its highwaymen, by the late nineteenth century the county was being civilized by the railway and, following it, wealthy people seeking refuge from London's polluted air. Azaleas were among the most successful plants for their gardens, and a thriving nursery business grew up to serve them.

ORIGIN
Eastern Europe, across
to the Caucasus

LONGEVITY
Long-lived shrub

SIZE
To 4 m (13 ft)

HABITAT
Woodland edges and
open woodland

COLOURS
This species: yellow;
hybrid derivatives come
in yellows, oranges,
creams, almost red

Rosa 'Rose de Rescht' (Rosaceae)

ROSE

Religious symbol

The rose is simply the most important flower in both the West and the Muslim world. Loaded with history and symbolism, and part of a large, complex group, it is for many simply the quintessential flower.

There are about 300 species of rose, and tens of thousands of cultivars. With a genetic centre of diversity in Central Asia, it is highly likely that the rose as we know it was first cultivated in ancient Persia, whence it travelled, mostly west. The ancient Greeks linked it to the goddess of love and beauty, Aphrodite. With Christianity, it became associated with the Virgin Mary and acquired deep symbolism; the rosary of Catholic prayer indeed refers to a garland of roses. Mystical currents in Catholicism have long linked roses and rose gardens to veneration of Our Lady. The rose was central to the Rosicrucians, a mysterious non-Christian philosophical movement of the seventeenth and eighteenth centuries. In Islam, it plays a central symbolic role in Sufism. Through the Mughal Empire this love of roses was transmitted to northern India, largely in culinary form; today rose-scented desserts, sweets and drinks are popular there, as in the Middle East.

The rose is also a national symbol for England, and appears in many other countries' iconographies. In the English civil conflict known as the Wars of the Roses (1455–85), the two sides had roses of different colours as emblems, the white rose of York and the red rose of Lancaster; the victor, King Henry VII, combined the two into the Tudor Rose as a symbol of unity. The United States made the flower its national symbol in 1986, and a red rose is a common symbol for European socialist or social democratic parties.

Roses became established as garden plants in Europe in the medieval period, but it was not until the late eighteenth century that serious breeding and selection began, led by French growers. Breeding increased almost exponentially in the nineteenth century, with more and more wild species being pulled into the gene pool; Chinese species were used particularly for the yellow flowers of some and the climbing habit of others. In the twentieth century rose-breeding and rose-growing diverged, being used for mass production on one hand, and for exquisite high-end garden plants on the other.

ORIGIN
Temperate climates across the northern hemisphere

LONGEVITY
Long-lived shrubs and climbers

SIZE
Wild species from 50 cm (20 in) to climbers potentially as big as a house

HABITAT
Woodland edges, scrub, steppe

COLOURS
Famously, everything except blue

Almost always vivid red at maturity, the rose 'hip' develops from the lower part of the flower.

*Elderberries are mature at
the end of the summer, and are
rapidly eaten by birds.*

Sambucus nigra (Adoxaceae)

Elder

Summer refresher

For much of European history the elder has had a powerful reputation, but one that veers wildly between negative and positive depending on time and place. In Britain, it tended to be a plant of witchcraft, not to be brought into the house or allowed to grow near it. This may be because of the unattractive smell of the leaves, or the fact that the timber is so weak; it somehow isn't a proper tree. In some places it was believed that the trees had a narcotic effect, making it dangerous to sleep near them. In Germany and Scandinavia, however, elder was the domain of Freya, goddess of home and garden, and was looked on favourably. The Romans and Greeks generally approved of it, too. Perhaps it was these pagan associations that made the Christian Church authorities turn against it. The myth arose that Judas had hanged himself on an elder branch (which probably would have broken if he had tried), and witches were said to be able to transform themselves into elders or even pieces of the wood. Elder also tended to be associated with death.

The berries can be used to make wine, or added to wine or port to deepen the colour. In the eighteenth century, when wines imported to northern Europe were too expensive for common people, elderberry wine was a good substitute, and arguably healthier. The flowers are extremely fragrant and traditionally were used to make a refreshing drink. Since the 1990s there has been a revival and a major commercialization of 'elderflower cordial'; many people also make their own 'elderflower champagne', using the natural yeast found on the flowers.

The elder was not grown in gardens until the eighteenth century, when the English landscape movement promoted the planting of trees and shrubs to create decorative wooded areas, or found other aesthetic or functional reasons for growing native specimens. In the twentieth century several new species were introduced into cultivation, and varieties selected for their golden-yellow or very dark foliage. Given the plant's growth habit, these are often treated as perennials, cut down to the ground every autumn and shooting up to flower in spring.

ORIGIN
Cool, temperate climates across the northern hemisphere

LONGEVITY
Long-lived shrub or small tree

SIZE
To 4 m (13 ft)

HABITAT
Moist, fertile soils, often disturbed ground around settlements

COLOUR
Creamy white

Trifolium repens (Fabaceae)

CLOVER

Horses' pasture

Clover is very familiar, although few of us actively grow it or appreciate it for what it is, or what it does for us. There are many plants that we call clover, most of them members of the genus *Trifolium*. Of these, we regularly see two spattered over grass: red clover (*T. pratense*) and our example, white. Since grass has no colourful flowers, our perception of grassland is often that of grass with its inevitable white or pink consorts. Native to the western half of Eurasia, clover travelled with pasture grasses all over the world. They usually grow together and have been encouraged to by European colonists because the combination feeds the cattle and other livestock central to the farming methods the settlers took with them. The clover helps to feed the grass, because, as a member of the pea family, it is able to 'fix' atmospheric nitrogen, turning it into nitrates; when the clover leaves and roots die, grasses can feed on the nitrates, and it is nitrates that make grass grow green, lush and nutritious. Together, grasses and clovers have transformed many grassland ecologies – often at great loss to local species.

The Romans knew how important clover was, not just to create healthy pasture, but also as food for livestock itself. For their army horses, they also grew the closely related alfalfa (*Medicago sativa*), which the Greeks had originally obtained from the Persians in about 500 BC – to all eyes, other than those of the botanist, yet another clover. Clovers have long been familiar as sources of nectar, and so are among the most important plants for beekeepers. Ironically, it is wild bumblebees that pollinate the flower most effectively, and indeed bumblebees have been introduced to many countries along with clover simply to ensure reliably regenerating crops.

The distinctive threefold division of the leaf is what defines a clover for most people. It is a shape that has long attracted attention as a source of symbolism or artistic representation. It is said that St Patrick used it to symbolize the Holy Trinity, hence it becoming a symbol for Ireland itself, although known by its Irish name, 'shamrock'. Perhaps this is also behind the love of the clover-derived trefoil seen again and again in European Gothic church architecture.

ORIGIN
Temperate climates from Central Asia westwards

LONGEVITY
Long-lived perennial

SIZE
To 15 cm (6 in)

HABITAT
Open situations as a component of grassland

COLOURS
White and pinks

A single clover flower illustrates the bilateral symmetry typical of the pea family at its most minimal.

*A camellia seed pod cracks open
to reveal large, oily seeds.*

Camellia 'Kuni-No-Hikari' (Theaceae)

CAMELLIA

Beloved of the samurai

A shrub or small tree that thrives in mild, wet climates, the camellia has become associated with lushness and decadence, as illustrated by Alexandre Dumas *fils'* novel *The Lady of the Camellias* (1848) which was adapted by Giuseppe Verdi into the opera *La traviata* (1853). This is not necessarily an association it has in its countries of origin; in China the camellia symbolizes fruitful marriage, and among the Japanese samurai a red camellia flower signified a noble death.

Although camellias are known from earlier records, cultivation in China did not start until the Three Kingdoms era (AD 220–280), and for many years it appears that only the wild types were grown (single, red). By the Tang dynasty (AD 618–907) other varieties are mentioned. It was the Japanese who really fell in love with the flower, with the elites growing them from the twelfth century and an explosion of varieties in the Edo period. Tokugawa Hidetada, the first Edo shogun, was particularly fond of them, and people who wished to ingratiate themselves with him would bring him new varieties. By 1868 there were some hundred different ones, exhibiting most of the features we know today. During the Edo period the persecuted Christian community in Kyushu used the flower to symbolize resurrection and perseverance. It was on this southern island that the camellia perhaps reached its high point. The Higo clan were notable growers of many plants, and their varieties are still acclaimed today for their bold central boss of stamens, a feature that is hidden in the more popular doubles.

With the increasing contact between East and West from the sixteenth century onwards, Western growers were desperate to obtain camellias perhaps more than any other plant. A trickle of introductions in the eighteenth century became a flood in the nineteenth, for these are easy plants to breed and propagate. Cold winter weather was no discouragement, since early glasshouses were given over to growing them across central Europe. The flower's peak was probably reached in the 1840s, before improved glasshouse heating and tropical plant-hunting brought in orchids to displace it as aristocratic status symbol. In the twentieth and twenty-first centuries it became widely established as a garden plant, with breeders making use of the many wild species to broaden the gene pool.

ORIGIN
East Asia

LONGEVITY
Long-lived shrub
or small tree

SIZE
To 6 m (20 ft)

HABITAT
Forest and
woodland edges

COLOURS
Reds, pinks, white; some
pale yellow cultivars

Convallaria majalis (Asparagaceae)

LILY OF THE VALLEY

Europe's sweetest scent

With a scent that is sweet, but not cloying, and rich, but not overwhelming, lily of the valley is perhaps the nearest that a cool temperate-climate flower produces to the heady scents usually associated with tropical flowers.

A locally common plant in the wild, it draws attention to itself because in time it can form enormous colonies, although the gardener who tries to grow it must be patient, as it can take several years to build momentum. It is unclear when lily of the valley was brought into cultivation (the first records are in the fifteenth century), but as a distinctive plant it often appears in mythology. An example is the legend from southeastern England that tells of it springing forth from the blood of a dragon slain by St Leonard.

So extensive are its colonies in undisturbed woodland that local festivities were sometimes organized to celebrate its flowering, especially since this often happened at the important Christian festival of Whitsuntide. One writer recorded that special huts were erected in the woods around the German city of Hanover, to serve drinks and tobacco to the many visitors who often stayed to dance away the evening.

As a cut flower, lily of the valley may be small but it lasts well. With the development of the railways, it began to be grown on a large scale for markets, both for its flowers and for the sale of the rhizomes for 'forcing' (growing in heat to produce early growth), a practice easily done on the windowsills of even the most humble home. Several areas of Germany specialized in its production, and the 25 hectares (62 acres) noted in 1914 around Frankfurt an der Oder were particularly noteworthy, with plants being exported all over Europe.

ORIGIN
Central and northern Europe; very closely related species in eastern Asia and eastern North America

LONGEVITY
Long-lived perennial

SIZE
Around 25 cm (10 in)

HABITAT
Woodland

COLOURS
White; some pale pink garden forms

*The berries of lily of the valley are
reddish or orange when mature.*

A seed capsule breaks open at the
top to release the small, very dark
seeds typical of Dianthus species.

Dianthus barbatus 'Vesuvio' (Caryophyllaceae)

SWEET WILLIAM

Grown for Henry VIII

The genus *Dianthus* includes many species that are happy to grow on the shallow, alkaline soils that are fatal to many others, and this has led them to flourish on and near stone walls. Since it was houses and castles with stone walls that defined the living spaces of the European aristocracy for many centuries, this brought them very close to the only people who had the time and resources to garden for several hundred years.

The flowers of many have 'pollen guides', markings that direct bees to the centre of the flower. These can make us think of eyes, which may explain the frequent mythological connections with eyes. The goddess Artemis, for example, is said to have shot out a young man's eyes with an arrow, and dianthus plants then sprang up from them as they lay on the ground.

Sweet william is known to have been in cultivation in France, England and Germany in the sixteenth century, possibly in monastery gardens. The English king Henry VIII is recorded as having had them planted in his new garden at Hampton Court, southwest of London, in 1533. It is known that the earliest recorded attempts at deliberate hybridization in the West were carried out on dianthus, by the nurseryman Thomas Fairchild in the early eighteenth century; the result was dubbed 'Fairchild's mule', a cross between this and a carnation. Concerns about whether this was blasphemy, on the grounds that it constituted interfering in God's creation, led him to pretend that it had happened naturally.

The sweet william diversified rapidly, and several colour forms were available by the early sixteenth century. It continued to be a popular plant until well into the twentieth century, when it rather lost its popularity, although forms with exceptional flower colours, such as our dusky example, have bucked that trend.

ORIGIN
Southern Europe, the Balkans and parts of the Middle East

LONGEVITY
Biennial

SIZE
Around 40 cm (16 in)

HABITAT
Shallow, calcareous soils

COLOURS
Pinks, reds, white

Gardenia jasminoides (Rubiaceae)

GARDENIA

Polynesian garland

The flowers of gardenias have a rich ivory colour and an extremely deep, exotic scent. The first recorded mention of this species is in manuscripts from China's Song dynasty, with illustrations of both single and double forms. It continued to appear in paintings and on porcelain in the succeeding centuries. Although this particular species is not native to the Pacific Islands, there are other closely related species there (this is a large genus), where it has a variety of ritual uses, such as in *leis*, the garlands popular among the Polynesian peoples of the region. Historically, these were reserved for the elite or used only for ritual purposes. Now, in Tahiti in particular, they have become, along with other highly scented flowers, a popular expression of local culture.

Brought to Europe at the end of the eighteenth century, having been obtained by British traders from gardens in southern China, the gardenia proved a difficult plant. It remained a rarity until the nineteenth-century greenhouse and heating-system revolution enabled nurseries to grow it on a large scale. The flowers then became the ultimate floral symbol of luxury for wealthy Americans and Europeans, perhaps even more so than orchids. Part of the reason was their expense and delicacy. Unlike orchids, gardenia plants are highly susceptible to anything outside their demanding requirements of heat, humidity and acidic soil; the flowers too are easily and quickly damaged by getting wet (which would never bother an orchid). Consequently, they became simply *the* flower for wealthy men to wear as a buttonhole.

G. jasminoides will grow outside in the American South, and the decision of Hattie McDaniel to wear such a floral symbol of privilege in her hair when she accepted the first Academy Award for an African-American film actor in 1940 (for her role in *Gone with the Wind*) was seen as a definite but subtle political gesture.

ORIGIN
South East Asia

LONGEVITY
Shrub

SIZE
To 3 m (10 ft)

HABITAT
An understorey shrub from evergreen tropical forests

COLOUR
Ivory white

Unfamiliar outside the tropics:
a gardenia seed pod.

The compound seed head of a hepatica.
Seeds can be very slow to germinate.

Hepatica nobilis (Ranunculaceae)

HEPATICA

Naturally variable

The genus *Hepatica* is named after the Ancient Greek for liver, on the supposed similarity of the leaf shape to that of the organ. Therefore, during medieval times in Europe, it was supposed to be beneficial for diseases of the liver, according to the pre-scientific Doctrine of Signatures. Found in humus-rich woodland soils from central Europe eastwards, it reaches a kind of genetic high point in Japan, with a species that throws up a remarkable number of colour variations (everything from blue-purple to bright pink) and flower forms, including doubles, with petals in a variety of shapes and, most unusually, in differing numbers. The level of variation is quite regional, however, and highest on the western side of the country; those on the Pacific Ocean side nearly all have white flowers. During the Edo period these were collected and sold in towns, but it was not until much later, during the twentieth century, that enthusiasts began to cultivate them properly, sowing seed to make the most of the natural variation and then propagating particularly good forms.

Towards the end of the Edo period, in the early nineteenth century, collecting good examples for the markets of Kyoto and Edo (present-day Tokyo) was an important seasonal business for country people in the regions where the plants were both common and variable. A number of books were published illustrating the range of different forms. Interest then lagged until the latter part of the twentieth century, when the plant became fashionable again. Prices soared, especially for unusual forms. As might be expected, wild populations began to suffer from the increased rate of collecting. However, the prices for special plants did at least drive the nursery trade towards naming and propagating good forms.

ORIGIN
From central Europe across Asia to Japan; related species in North America

LONGEVITY
Perennial

SIZE
About 20 cm (8 in)

HABITAT
Woodland

COLOURS
Blue-purples, pinks, reds, white, very occasional yellows

Hibiscus rosa-sinensis (Malvaceae)

HIBISCUS

Everlasting flower of Korea

This is a genus of more than 700 species, of which three dominate cultivation: the relatively cold-hardy small tree *H. syriacus*, the tropical shrub *H. rosa-sinensis* and the herbaceous *H. moscheutos*. The woody species are noted for the tough fibres in their stems, and so several have been used for making robes and cloth in traditional cultures, while one, kenaf (*H. cannabinus*), has been used in paper-making. *H. sabdariffa*, known as Roselle, is widely eaten as a vegetable in Africa and Asia, while around the Caribbean, the flowers are used to prepare a refreshing drink.

H. syriacus has been grown since at least the seventh century in Korea, where it is known as 'everlasting flower' and embodies the survival instincts of the Korean people and their culture; the country has even been known historically as Hibiscus Land. The five petals of the flower are a common element in Korean iconography.

As a garden flower hibiscus has the special quality of being one of the very few summer-flowering large shrubs in temperate regions. The range of commercially available varieties is limited, but flower colours range from mauve through pink to white. As with most members of the Malvaceae, hibiscus flowers are short-lived and can be enjoyed only on the living plant. The breeding of new varieties, with flowers in different colours, and occasional doubles, has played a major role wherever they are grown. This is particularly the case with our example, the tropical *H. rosa-sinensis*, which is one of those plants that seems especially rewarding for amateur plant-breeders; the range of colours and patterns, and to some extent flower forms, being developed is truly extraordinary, and arguably some of the most dramatic of any plant species. An international society keeps enthusiasts in touch, and registers their new creations.

Among the herbaceous species, *H. moscheutos* has long been popular as a garden flower in the United States. It and several related wetland plants thrive in hot, humid summers, surprising with their outlandishly large flowers. It is naturally pink, but shades from deep red to pure white have been produced by breeders.

ORIGIN
Widely distributed in humid tropical and subtropical climates; a few species in temperate Asia

LONGEVITY
Varies from short-lived perennials to moderately long-lived

SIZE
Large perennials, woody shrubs and small trees

HABITAT
Generally fertile soils in forest edge and disturbed habitats

COLOURS
Overwhelmingly pinks and reds; blue pigments absent

*The central structure of the
hibiscus flower is quite unlike that
of any other, combining a means
of projecting both female and male
parts away from the petals.*

*The central structure of a
magnolia flower, with stamens
below and female styles above.*

Magnolia x *soulangeana* (Magnoliaceae)

MAGNOLIA

Aristocrat of spring

Magnolias are popular spring-flowering trees, but they have taken a long time to move down the social ladder. Most are large trees, descendants of one of the more ancient botanical lineages, having evolved during the Cretaceous period, before the evolution of bees; it is thought the earliest magnolias were pollinated instead by beetles. First cultivated in China, during the Tang dynasty (often as a bonsai), the yulan, *M. denudata*, was the first Asian species to make its way to Europe, during the eighteenth century. Others followed, but all were substantial trees that took a long time – often 25 years or more – from seed to flowering, while propagation from cuttings was problematic. They tended to be grown only in the large gardens of the wealthy.

A breakthrough came in 1820, when a retired cavalry officer in Napoleon's army, Étienne Soulange-Bodin, made a cross between *M. denudata* and *M. liliiflora*, and was taken aback by its flowering only six years later. The cross has been remade many times, and so we now have a flurry of *M.* x *soulangeana* varieties, in many different colours and with flowers in a range of sizes. 'Grace McDade', for example, bred in the American South, has flowers up to 35 cm (14 in) across. Conveniently, they are relatively small trees, big enough to make an impact in the suburban garden but not getting so large as to cause problems. The opening up of Japan brought the small-growing *M. stellata* to the West, initially to the United States, and to Europe via the most innovative of British nurseries, Veitch of Exeter. Together *soulangeana* and *stellata* democratized the magnolia. Advances in grafting techniques and new hybrids have further increased the range that is available to flower at young ages and obtainable at reasonable prices.

ORIGIN
A plant of hybrid origin whose parents originate in southern China

LONGEVITY
Long-lived tree

SIZE
6 m (20 ft); some varieties possibly taller

HABITAT
The parents are trees of mature forest

COLOURS
White and many shades of pink

Prunus x *yedoensis* 'Somei-Yoshino' (Rosaceae)

JAPANESE CHERRY
Sacred to Buddhists

Japan is famous for its cherry blossom (*sakura*), the first sign of spring, and the *hanami* parties that are held to celebrate it, when mats are spread out beneath the trees and families and work groups sit down to picnic. The event starts in the south and spreads northwards, so that Tokyo and everywhere south is celebrating by mid-March, but Hokkaido in the north does not join in until the end of April. The idea of holding cherry-blossom parties started only in the Edo period, although the roots of cherry-growing and appreciation are much older. The first records are from the seventh century, when trees were grown around Buddhist temples, with a particular link to the mystical Shugendo sect, who selected particularly good trees on journeys into the mountains. Selected forms were named and propagated at this time.

For centuries cherries played second fiddle to an even earlier-flowering tree, *P. mume*, known as *ume*. This was the case until the twelfth century, when it is thought grafting was developed, enabling the propagation of selected varieties and ensuring complete reliability. Many cultivars were developed in the early nineteenth century, including our example, which probably has parent species from China and Korea. Its growing popularity towards the end of the century may have been one of the reasons for the loss of a great many other cultivars. Indeed, nearly all Japanese cherry blossom is just this one cultivar. The uniformity of cherry blossom from one end of the country to another is one of its appealing aspects, as is its transience, which is an immensely important quality in Japanese culture. Indeed, kamikaze pilots in World War II were also known as *sakura*, young men who had a similarly brief flowering.

ORIGIN
Parents of this cultivar are from East Asia

LONGEVITY
Short-lived tree, maximum 100 years

SIZE
To 9 m (30 ft)

HABITAT
Parents are found in open woodland and at forest edge

COLOUR
Pale pink

Cherry fruit, each with a large,
hard stone containing the seed.

The immature seed capsule of a violet, which typically bursts open on three sides when mature, to release seed.

Viola odorata (Violaceae)

SWEET VIOLET

Fugitive fragrance

The perfume of violets is truly evanescent – smell it a second time, and it has gone! This is because ionone, the compound that is responsible for the scent, has an anaesthetic effect on the organs of the olfactory system. Known since classical times, it was possibly first cultivated by the Arabs, and passed on to the Ottomans and to Europeans during the medieval period. Varieties in different colours, as well as doubles, are known from the sixteenth century. Large-scale cultivation began in the early eighteenth century in France, especially around the perfumery city of Grasse in Provence. The French also used the flowers in confectionery, cakes, preserves and salads. The area around Toulouse became a centre for the production of violet culinary products during the nineteenth century, and to some extent it still is, while the flower is a symbol of the city.

Violets are rather short-lived as cut flowers, but in the romance-obsessed late nineteenth century this was no bar to extensive commercialization. As with many plant products, it was the railway that facilitated their production and sale. And being small helped, since transport costs would have been very low. Areas near towns in the southwest of England often attracted growers, as the mild climate there allowed early production, and the flowers were only an overnight train journey from London. In the United States they were apparently grown only around large population centres; for example, New York's were all grown around the town of Rhinebeck in the Hudson Valley, which once had more than 100 businesses, mostly run part-time, growing the flowers. As quite a vigorous spreader (in the right place), violets have often survived in hedgerows around the former nurseries.

ORIGIN
Much of Europe
and western Asia

LONGEVITY
Long-lived perennial

SIZE
To 15 cm (6 in)

HABITAT
Woodland edge

COLOURS
Violet; also purple, blue,
pink and white forms

Wisteria sinensis (Fabaceae)

WISTERIA

Favourite of the Fujiwara

Few flowers have attracted the attention of artists as much as the wisteria. This climbing plant that is truly impossible to treat as a cut flower has enjoyed another life, as the subject of paintings and prints, in both East and West, high art and kitsch. Wisteria does not flower for very long, but it makes up for this with a vast quantity of spectacular flowers in shades of violet-blue, its unusual pendant quality marking it out as special. Its short season of glory gives it great importance for the Japanese, whose culture enjoys fleeting beauty; its name there, *fuji*, derived from the word 'to blow down', refers not to the plant in bloom, but to petals scattered on the ground.

The two species most commonly grown are *W. sinensis* and *W. floribunda*, associated with China and Japan respectively. The plant was almost certainly grown in Tang dynasty China, while during the Heian period (AD 794–1185) in Japan, wisteria-viewing parties became popular, especially since the flowers were a symbol of the ruling Fujiwara clan. Clambering naturally to the top of trees with their powerful twining stems, plants were also sometimes trained over buildings. The use of pergolas – quite possibly invented to support wisteria – as supporting structures for wisteria appears to have become popular in the late seventeenth century, and famous ones attracted numerous visitors. The world's largest wisterias are in parks in Saitama prefecture, near Tokyo. The largest North American example is at Sierra Madre, California, dating from 1894; an annual festival has been held in its honour since 1918.

Introduced to the West by the plant-hunter Robert Fortune in the mid-nineteenth century, wisteria rapidly became popular, and was displayed particularly well through training on horizontal wires attached to buildings. It is a hard flower to improve on, and variety selection has focused largely on the colour of the flowers and length of the racemes. Meanwhile the pergola took off during the early twentieth century as part of the Western Arts and Crafts garden, supporting increasing numbers of other climbers.

ORIGIN
The two most widely cultivated species are from the Far East; others are found in North America; all temperate summer rainfall climates

LONGEVITY
Can live for several centuries

SIZE
Lianas (vines), growing up to 150 m (490 ft) long and capable of reaching to at least the height of a five-storey building

HABITAT
Forests and forest edges

COLOURS
Blue-purples

The seed pod makes it immediately clear that this is a member of the pea family.

The seed pod of a snapdragon (seen here in its immature state) releases tiny seeds through a small opening when mature.

Antirrhinum majus (Plantaginaceae)

SNAPDRAGON

Laboratory subject

Popular as garden plants, these free-flowering perennials are named in most languages after something with jaws: *boca-de-leão* and *großes Löwenmaul* (Portuguese and German respectively – lion's mouth), *muflier* (French – muzzle). A classic bumblebee plant, its flowers will open only for heavier wild bees, not the comparatively light domestic honey bees.

The snapdragon has been recorded since the medieval period, quite probably because it is one of those plants that flourish on old walls, and therefore brings itself right into urban areas. It was in cultivation by the sixteenth century, and by the early nineteenth there were several colours, as well as doubles. At that time the nursery and seed trade did a lot of breeding to diversify the range of colours further, including bi-coloured and striped varieties, and to bring variations in size and plant character. Garden books of the period were almost apologetic about discussing how to grow something that very often grew itself on ruined buildings. The ones we grow today rapidly reach 40–50 centimetres (16–20 inches), but tend to stay around that height. Some nineteenth-century varieties, however, grew to over 1 metre (3 feet), even 2 metres (6 feet) (although taking longer about doing it), and formed spectacular mounds of flower all summer long. Modern breeding has focused on producing predictable, smaller plants that can be used in bedding schemes and window boxes.

The snapdragon has a life beyond the flower garden: in the laboratory. Widely used as a modern organism since the early years of genetics, it is employed in the study of inheritance, plant development and physiology. It is highly variable and easy to grow; its genetics are relatively straightforward; and it can be readily both self-pollinated and cross-pollinated.

ORIGIN
An arc of the northern Mediterranean region between Morocco and Syria

LONGEVITY
Short-lived perennial

SIZE
Generally about 50 cm (20 in); different seed strains will vary

HABITAT
Rocky places

COLOURS
Reds, pinks, white, yellows, pale oranges; no blues or purples

Aquilegia vulgaris (Ranunculaceae)

COLUMBINE

Shaped like a dove

The common name is derived from the Latin for dove, from a supposed likeness of the petals to the shape of that bird – the beak being the upwardly projecting nectary. The Latin name, however, recalls the eagle. Folk names often liken these flowers to hats, and indeed in English they have sometimes been known as 'granny's bonnet', recalling a style of headgear long forgotten.

Known in medieval times as a medicinal herb, columbine is actually mildly toxic, and by the eighteenth century it had been dropped by apothecaries and made it to the flower garden instead. These easy-to-grow flowers are particularly rewarding because they show a lot of variation, which can often increase from one seed-sowing to another, rather than new plants gradually resembling the wild species more and more, as is the case with most garden plants. The seventeenth-century English botanist John Parkinson noted also that, unusually, the doubles gave as good a quantity of seed as the singles. Popular as garden plants by the nineteenth century, they became particularly associated with cottage gardens, that is to say, the homes of the rural poor, a good indication of how easy to grow they were, but also how rewarding.

By the late nineteenth century the very wide genetic diversity of the American range of species was introduced, and many hybrids made. Indeed, the plants are notoriously 'promiscuous', and it is very difficult to keep seed strains pure. The western American species in particular proved very useful for breeders, with their large flowers and dramatically long spurs. A notable twentieth-century development has been breeding with smaller, high-altitude Japanese species, to produce miniature plants for rockeries and containers.

ORIGIN
This species: northern and central Europe

LONGEVITY
Short-lived perennial

SIZE
Generally about
1 m (3 ft)

HABITAT
Woodland edges and glades

COLOURS
Original species is dark blue-purple; cultivated forms may be pink, white or very dark purple

Divided into five compartments,
the seed heads of columbines split
open from one end to the other.

The tiny seed of Campanula
*is released through small holes
in the top of the capsule.*

Campanula pyramidalis (Campanulaceae)

BELLFLOWER

Usually blue, always bells

Members of the genus *Campanula* are characteristic of calcareous soils, including a good many Mediterranean species that seem happy growing in cracks in solid limestone. A few have been popular with gardeners since the Middle Ages. One, *C. rapunculus* (rampion), was grown as a root vegetable across Europe for centuries, and is the plant that is central to the fairy tale of Rapunzel. *C. medium* has the longest history in cultivation, and its English common name 'Canterbury bells' commemorates the small bells that hung on the horses ridden by pilgrims to Canterbury. Our example, *C. pyramidalis*, appeared in the sixteenth century but is now rarely seen. One of many with a short life span, this is usually grown as a biennial, its rosette of leaves sprouting a magnificent spire of flower to 1.5 m (4 ft) in its second year. It was known as the 'chimney bellflower' since from the seventeenth century to the nineteenth it was used as a decoration for the empty fireplaces of summer. Another common use for it was in the formal parterre plantings typical of the Baroque era, where single plants were surrounded by coloured gravel and hedges of clipped box.

The period around the turn of the twentieth century was a heyday of bellflower introduction, since this was the era of the rock garden, and many bellflowers are happiest there. The diminutive but easy *C. cochlearifolia* charms many on rockeries or in trough gardens, while others, most notably *C. poscharskyana*, have enormously running underground stems, enabling them to run through the mortar of old walls to reappear (delightfully or annoyingly, depending on circumstances) many metres from where they were planted.

ORIGIN
Northern hemisphere, centred on southeastern Europe. This species: Italy to the Balkans

LONGEVITY
Biennial, short-lived perennial or true perennial

SIZE
10 cm – 1.5 m (4 in – 4 ft). This species: 1.5 m (4 ft)

HABITAT
Mostly calcareous meadows, woodland glades; some only on rockfaces

COLOURS
Blue-purples; a few white

Canna 'Picasso' (Cannaceae)

CANNA LILY

Native American sustenance

Plants of the tropical Americas, cannas are known to have been an important food source for thousands of years, and possibly one of the oldest domesticated plants in the New World. Over time, however, they were displaced by more productive sources of carbohydrate.

Cannas appeared in Europe in the mid-sixteenth century, probably via contact with the Spanish Empire. By the end of the eighteenth century German writers were promoting them as an exotic summer bedding plant that could be easily kept over winter as dormant tubers in a frost-free storeroom. The first breeding was by the French Consul to Chile, who worked on them during his retirement in the 1850s. These plants were initially very tall (3 m/10 ft) and grown for their foliage, and it was a later generation of French nurseries that made more of the flowers and, perhaps crucially, produced shorter plants. Nurseries near Porte de la Muette in Paris were famous for producing tens of thousands of specimens for planting out in parks throughout the city. The fashion spread, and by the end of the century cannas were being widely grown in the gardens of the wealthy and parks all over Europe and, increasingly, in the European colonies. A key event in popularizing them may have been the World's Columbian Exposition in Chicago in 1893, where there was a bed of them 300 m (984 ft) long; after this American breeders began to produce their own varieties in great quantities.

Cannas have become one of the most important decorative plants of the tropics, as well as being popular anywhere with warm summers. Our example probably dates from mid-twentieth-century France, and is notable for its dramatically spotted flowers, a characteristic that has kept this variety popular while many older ones have been replaced by the results of more recent breeding.

ORIGIN
Tropical South America

LONGEVITY
Long-lived perennial

SIZE
To 3 m (10 ft); garden forms are smaller

HABITAT
Wetlands, riverbanks

COLOURS
Reds to yellows; some pink species; no blues

The tuber of a canna,
with shoot emerging.

The papery seed pod has burst open, releasing its seeds, which are usually distributed by ants.

Cyclamen persicum (Primulaceae)

CYCLAMEN
Beloved of boars

Long used as a medicinal herb, with the usual lack of concrete evidence for any beneficial effects, cyclamen seems to have been most popularly used in childbirth. So potent was it supposed to be that John Gerard, writing in sixteenth-century England, built a fence around those in his garden, in case a pregnant woman should come too close and suffer a miscarriage as a consequence. The round corms have a poor flavour, but are dug up by wild boars or domestic swine let loose in the woods. This explains the old English name 'sowbread', possibly just a translation from the French *pain de porceau*, since the plants are not native to Britain.

The western species (as opposed to those from the eastern Mediterranean) – the late summer-flowering *C. hederifolium*, which can carpet the ground with pink or white flowers in old gardens, and *C. purpurascens*, a less tractable plant – were the first two to be written about, during the post-medieval period. *C. persicum* arrived in western Europe during the seventeenth century, but its lack of hardiness kept it a sheltered novelty.

The development of greenhouses and the mass market in garden plants of the nineteenth century gave *C. persicum* a new life as one of the most popular flowering pot plants. A mutation resulted in chromosome doubling and consequently much larger and more vigorous plants, leading to the development of a great number of varieties for winter sale. The great British writer of domestic advice Mrs Isabella Beeton declared them 'most beautiful, graceful and ladylike; so easily cultivated withal'. Ideal for the Christmas market, the plant has never looked back, with breeding producing an ever-increasing range of flower colours and novelties such as picotees, where the petals are edged in a different colour, fringed petals, and leaves with silvery markings. French and German companies tend to lead in the breeding of cyclamen. Since the last decade of the twentieth century there has been, as with so many plants, a stepping back from the 'bigger is better' mentality of previous generations, with ranges of smaller plants being bred, many with attractively marbled leaves. These are increasingly used in bedding schemes in regions where winters are relatively mild.

ORIGIN
Mediterranean basin from Italy eastwards to the Caucasus

LONGEVITY
Long-lived perennial

SIZE
20 cm (8 in); cultivated forms at least twice this

HABITAT
Woodland and woodland edge

COLOURS
Pinks; occasionally white

Dahlia 'Bishop of Llandaff' (Asteraceae)

DAHLIA

Czech symbol

Few flowers are as varied as the dahlia. Since soon after its introduction to Europe from Mexico in 1803, breeders have produced plants in a bewildering range of forms and colours (but, as so often, no blues). Easy to grow and to overwinter as tubers inside, the dahlia has been perhaps the ultimate combination of luxury and democracy, enabling everyone with a garden to have big, bold, exotic flowers.

Initially grown in public parks, particularly in Germany, the dahlia made big advances quickly, with 103 varieties by 1817 and a specialist nursery in Bad Köstritz, Thuringia, in the 1810s; the town is still a centre for dahlia-growing and there is even a dahlia museum. The flower was also popular in the Austro-Hungarian Empire, and the Czech nationalist and writer Božena Němcová got herself crowned 'Dahlia Queen' a few days after her wedding in 1837 at the 'dahlia ball' organized by the recently formed Czech Dahlia Society. The society then became a front for the revival of the Czech language and for political agitation. The composer Bedřich Smetana also composed a dahlia polka.

In Britain, the National Dahlia Society was formed in 1870, and the flower went on to become a real favourite of working-class gardeners, one reason being that its late summer season and multiplicity of distinct categories (based on flower shape) fitted in with the competitive vegetable shows that are a big part of Britain's gardening culture. By the latter part of the twentieth century a certain snobbery had set in, with middle-class gardeners turning up their noses at the plant – until, that is, this particular variety for some reason became socially acceptable. When the well-connected writer Christopher Lloyd began growing them in the 1980s in his garden at Great Dixter, East Sussex, they were back 'in' again.

ORIGIN
Central America

LONGEVITY
Perennial

SIZE
Cultivated forms to
2 m (6 ft); some wild
ones twice that

HABITAT
Woodland edges,
roadsides

COLOURS
Originally purples
and oranges; in
cultivation almost
anything except blue

*The seed head of a dahlia keeps the
old bracts that enclosed the flower.*

*A disc floret with its single,
very short petal.*

Echinacea purpurea (Asteraceae)

CONEFLOWER

Healing herb

The coneflower is one of those rare plants that enjoy a reputation as both healing herb and popular garden flower. Much used medicinally by Native Americans, often for common cold-type ailments, it became popular in folk medicine during the nineteenth century, and then again in the late twentieth century, supposedly as a booster to the immune system, although hard evidence has always been difficult to come by. Trials of its efficacy have been complicated by the fact that there are several species and it has never been clear which is supposed to be the most beneficial.

Although it has been in cultivation since the late eighteenth century, the coneflower only really became popular towards the end of the twentieth, largely because it became the poster boy for the burgeoning native-plant movement. The gardening public loves daisies, it seems, and since this was also a good flower for pollinators, and a source of seed for wild birds in winter, there seemed many more reasons for growing it in the increasingly ecologically aware years at the end of the century. The flower began to appear with increasing regularity on the posters, letterheads and websites of anyone or any organization promoting or selling native plants and the concept of garden biodiversity. Breeders, however, could not leave it alone and made crosses with other species, notably the yellow *E. paradoxa*, producing a range of yellow, orange and apricot hybrids with great commercial potential but an even shorter life span than the already short-lived *E. purpurea*. In warm-summer climates, however, the species easily self-seed, enabling sustainable populations to build up, at least as long as the gardener is not too ruthless in weeding.

ORIGIN
Eastern half of the United States

LONGEVITY
Short-lived perennial

SIZE
About 1 m (3 ft)

HABITAT
Prairie, woodland edge

COLOURS
Pink; cultivated forms include reds, oranges, yellows and white

Gentiana acaulis (Gentianaceae)

GENTIAN

Alpine decoration

Getting out of the cable car into the fresh air of the European Alps during early to mid summer, we are likely to be greeted by an amazing carpet of colourful wildflowers, amongst which the intense blue of this species is often the most prominent. It has indeed become an icon of the alpine regions, a status shared with the grey-white flowers of the edelweiss (see page 189), appearing on a vast range of merchandise, of which the boxes of Swiss chocolates are probably the best known, but also the folksy products with which the peoples of the alpine regions celebrate their distinctive cultures: painted, sculpted, printed or embroidered gentians appear on furniture, clocks, clothing, table linen and more or less anything that can be customised.

Historically, however, this was not the important gentian. Pride of place goes instead to a plant that looks so different that we find it hard to believe it is so closely-related, the metre high, yellow-flowered *Gentiana lutea*, whose roots are used in a range of alcoholic drinks, including Suze, Underberg and Angostura Bitters. The flavour is intensely bitter, and for centuries has been regarded as being curative of a great many complaints.

G. acaulis can be grown in the lowlands, as a rock garden plant. However it is one of those plants that seems to flourish in one person's garden, but refuses to grow in the next. Or it grows to form great mats, but then fails to flower. Much more tractable are the sprawling species of the Himalayas such as *G. sino-ornata*, but only in cool summer climates such as the west of Scotland. These can be so vigorous as to be regarded as almost weeds in some gardens.

ORIGIN
Mountains of central Europe

LONGEVITY
Long-lived perennial

SIZE
No more than 15 cm (6 in)

HABITAT
Mostly above the treeline

COLOURS
This species: blue; others may be white, yellow or purple

Gentians have distinctive seed
capsules that, when ripe, split
into two lobes at the tip.

The bristly outer coat of the
seed capsule attaches to animals
to get itself transported.

Myosotis arvensis (Boraginaceae)

FORGET-ME-NOT

Worn by lovers

The forget-me-not – its common name recalling an old European tradition that lovers who wore the flower would never be forgotten – is a common annual of waste ground and fields. The very similar but more robust *M. sylvatica* is a wild plant of light shade, and the ancestor of several varieties grown as garden plants since the early nineteenth century. As is so often the case, it is not always clear what species old traditions are in fact referring to, and here it is possible that the somewhat larger water forget-me-not, *M. palustris*, is being referred to, especially since some of the legends and stories involving the plant make reference to ardent lovers falling into rivers and being swept downstream begging not to be forgotten. It was used as an emblem by the English king Henry IV, while in Germany it became popular during the nineteenth century as a grave flower. It was also there that the late eighteenth-century botanist Christian Konrad Sprengel used its blooms as an example of how bees and other pollinators were attracted to flowers, leading him to advance the theory that flowers are in fact sexual organs. Today we might be astonished that this fact should not be obvious, but previous generations had no idea that this was the case.

Very easy to grow from seed sown in summer, the plants overwinter to flower in spring, a useful characteristic for cool climates. They have long been popular in municipal bedding schemes with tulips. As a garden plant they never go away, since the seeds cling to clothing and to the fur of small dogs and cats, and so are distributed far and wide.

ORIGIN
Northern and central Europe

LONGEVITY
Annual or short-lived perennial

SIZE
Rarely more than 40 cm (16 in)

HABITAT
Woodland and field edges, particularly where soil has been disturbed

COLOUR
Clear blue

Passiflora quadrangularis (Passifloraceae)

PASSION FLOWER

Christian symbol

Named for the Passion – the martyrdom of Christ – this large genus of American climbing plants has been loaded with symbolism. Discovered (in the seventeenth century) towards the end of the time when it was supposed that all living things were made for human benefit, and could therefore communicate hidden messages to humanity, the complex and botanically very distinctive flowers were seen as deeply symbolic by the Spanish Catholic conquistadores of the New World: the ten petals and sepals representing the ten faithful Apostles; the radial filaments the Crown of Thorns; the ovary the Holy Grail; the three stigmas the three nails; the five anthers the five wounds; and so on. Non-Catholic cultures, however, tend to liken the flowers to the face of a clock, and the plant is known as 'clock flower' in a number of languages. Such complex floral structures are actually all about the encouragement and manipulation of pollinators: bees but also hummingbirds or bats, depending on the species of passion flower.

The genus is the source of a commercially important fruit (*P. edulis*, the passion fruit) and many ornamental species, some of which are widely grown in warmer-climate regions; one, *P. caerulea*, is frost-hardy. Nearly all are rampant climbers, and so can be useful for screening or shading. The flowers of our example, known as the giant granadilla, are particularly impressive, especially the prominent filaments. The fruit is the largest of the genus, up to 30 cm (12 in) long; it is grown commercially for local markets, and named cultivars are available in some South American countries and in Indonesia. A large and very vigorous plant, it can rapidly smother trees and buildings.

ORIGIN
Tropical America;
exact origin unknown

LONGEVITY
Long-lived woody
climber

SIZE
Allegedly to 40 m
(130 ft)!

HABITAT
Forests and forest edges

COLOURS
This species: purple;
others reds, pinks,
lavenders and
yellow-greens

Species of Passiflora *typically have juicy fruit, containing a number of hard seeds.*

*The seeds of marigolds are thin and
quite sharp, with a papery attachment.*

Tagetes patula 'Honeycomb' (Asteraceae)

FRENCH MARIGOLD

Introduced by an emperor

Possibly the most popular 'bedding plant' in the world, this is a toughie that can be raised under cover in early spring, dug up and sold in twists of newspaper on market stalls in the poorest of countries, to adjust rapidly to new surroundings and then go to flower with outrageously bold yellow or orange blooms all summer. It has no more to do with France, however, than the very similar 'African' marigold (*T. erecta*) has with Africa.

Quite possibly grown by the Aztecs, the *cempoalxóchitl* (as it is known in Nahuatl) has been identified in sculptures in various locations in Mexico, and was brought to Europe by the Spanish in the sixteenth century. Over the next century it was spread across Europe, and it acquired its English name by being brought to England by French Protestant refugees. The African marigold got its name from the fact that although the Spanish introduced it, they did not grow it on any scale until some crusaders saw it flourishing in North Africa during a brief raid by Emperor Charles V. The marigolds were not universally liked, for the scent of many varieties is strong and not especially pleasant. From the mid-nineteenth century breeders went to work in earnest with the plant, focusing on doubles and compact plant size. Today's dwarfs, with their outsize flower heads, are a far cry from the early forms, which had single flowers and could grow to 1.5 metres (4 feet).

Its easy growth has meant that this plant has rather pushed others out of the way, even usurping 'marigold' from *Calendula officinalis* (see page 17) in the popular imagination. In Hindu temples, where flowers are needed in industrial quantities, it has inevitably displaced that and other less productive species as the raw material for garlands and offerings.

ORIGIN
Mexico and Central America; Caribbean

LONGEVITY
Annual or short-lived perennial

SIZE
Wild plant to 1.5 m (4 ft); garden forms often under 30 cm (12 in)

HABITAT
Open places in forests; disturbed ground

COLOURS
Yellow-orange, with brown-red tones in modern varieties

Vinca major (Apocynaceae)

PERIWINKLE

Medieval evergreen

Technically a shrub, rather than a perennial, and evergreen, periwinkles may be low-growing but they are tough survivors. Their habit of projecting stems out over other plants to root on the other side gives them a competitive advantage and enables them to survive and spread without the supporting help of the gardener. In those areas where they are garden plants rather than natives, they typically indicate the sites of long-abandoned gardens or where they were dumped as waste from gardens, often decades ago. The downside of this is that in some regions they can smother local plants and acquire that dubious distinction of being 'invasive aliens'.

Used by the Romans for making wreaths (the stems are ideal for winding around each other), the plant has probably been in cultivation in central Europe and England since the medieval period; it was mentioned by the fourteenth-century English poet Geoffrey Chaucer. It would have been one of the very few evergreens with which people of the period were familiar, and indeed its German name, *Immergrün*, simply means 'evergreen'. In England it was used to make crowns for criminals or those accused of treachery to wear on their way to execution; in Italy its macabre associations were more to do with its use for garlanding dead children. Being evergreen and willing to grow in shade soon made it a popular plant for growing on graves, particularly the smaller species *V. minor*, and it is still popular in cemeteries in Germany. A double appeared in Flanders in the late sixteenth century, and from the eighteenth century onwards numerous colour and variegated forms from both this and *V. minor* were developed as cultivars.

ORIGIN
Southern Europe

LONGEVITY
Long-lived shrub

SIZE
To 30 cm (12 in), but crucially spreading sideways as much as 50 cm (20 in) a year

HABITAT
Woodland

COLOURS
Blues; cultivars offer purples, red-purples and white

The neatly furled flower
bud of a periwinkle.

A zinnia bud showing the
distinctive 'involucral bracts'
that surround the base.

Zinnia elegans 'Benary's Giant' (Asteraceae)

ZINNIA

Colours from the Aztecs

Named after an eighteenth-century German botanist, Johann Gottfried Zinn, zinnias' popularity has gone up and down since their introduction to Europe by the Spanish from their Mexican colony in the late eighteenth century. Not as easy or resilient as the *Tagetes* marigolds (see page 113) that come from the same area, they have a far wider colour range, and indeed provide us with some of the most intense colours available from any garden plant. It is highly probable that zinnias were grown by the Aztecs, and the range of colours the plants showed in Europe is probably a legacy of this culture's variety selection; as early as 1842 a German book listed an impressive colour range. The flowers are noted for having both bright pink and bright yellow forms, a colour combination that is popular today in Mexico, although seen as a 'clash' by many Europeans.

Zinnias became very popular in the nineteenth century, but suffered something of a decline in the late twentieth. Today, with the growing sophistication of summer bedding schemes and the demands of the floristry industry, the plant seems to be becoming more fashionable again. As with much creative breeding for floristry, growers in Japan seem to be playing a crucial role.

Our example is one bred by Benary, a German seed company that was founded in 1843 by Ernst Benary and became one of the most important breeders of ornamental plants in the world. It established the city of Erfurt as Germany's garden capital, and continued to be run by the family. It had to relocate to West Germany during the Communist period, but Erfurt has continued as a major horticultural centre. The company is still an important one in the commercial production of flower seeds, and its zinnias are noted for their magnificent size and colours, which range from the intense to the subtle and surprising.

ORIGIN
Southern United States, Mexico, Central America

LONGEVITY
Annual

SIZE
To 90 cm (35 in)

HABITAT
Open places, waste ground, roadsides, woodland edges

COLOURS
Reds, oranges, yellows, pinks, purples, greens, a few whites

Galanthus nivalis (Amaryllidaceae)

SNOWDROP

Emblem of chastity

Universally admired in temperate-zone gardens for its ability to sprout as the first flower of the year, the snowdrop has an early history that is obscure. It does not appear to have reached northern Europe until the sixteenth century, but its use as an altar flower for the Virgin Mary suggests that this must have been common in southern Europe, where it originates. Once introduced, it spreads rapidly, since it regenerates very easily from its little bulbs whenever the soil is disturbed, potentially forming huge colonies. In the Netherlands it is one of the *stinzenplanten*, literally 'stone house plants', that became associated with the estates of the wealthy, who often had some sheltering woodland around their gardens, where small early-flowering bulbs could easily spread over the centuries.

By the Victorian period, the snowdrop was well established up and down Britain, and as an emblem had strong associations with chastity. In particular it was used in campaigns led by the churches to police the behaviour of the many working-class girls who, newly liberated from home and farm, were working in the mills of the country's burgeoning industrial cities, and having too good a time for the taste of some. Very similar to the many temperance societies of the time, a Snowdrop Club was established, with a magazine of moral tales, *The Snowdrop*, and even a Snowdrop Band, which operated around 1890.

This is simply one of some twenty species, many of which are also popular in cultivation. There are also a great many cultivars. Most of those are relatively minor natural genetic variations, but, because of the plant's ability to clone itself, are easily reproduced, although building up numbers takes time; new varieties can therefore be very expensive indeed, and sometimes even the victims of theft. Admiring the plants at 'Snowdrop Lunches' has become something of a new tradition among elite British gardeners as a start to the gardening season in January.

ORIGIN
France and Italy, eastwards to Ukraine; naturalized in northern Europe

LONGEVITY
Long-lived perennial

SIZE
About 20 cm (8 in)

HABITAT
Woodland

COLOURS
White; very occasionally very pale yellow

An immature snowdrop seed pod.
The seeds are typically
released in late spring.

The seed heads with their distinctive
'horns'. Seed is released suddenly,
making collection difficult.

Helleborus x *hybridus* (Ranunculaceae)

HELLEBORE

Perilous medicine

Hellebores are one of those flowers that, if performing in summer, would be largely ignored, since the flowers are generally dull versions of brighter colours. However, they flower very early in the year, which makes them as much appreciated by gardeners as by bees.

The plants have long been used in herbal medicine and witchcraft, although they are quite toxic, so medical use must have been fraught with danger. The ancient Greeks ritually circled the plant with a sword before digging it up. Their early flowering made hellebores popular as garden plants from the sixteenth century onwards, especially the usually white-flowering *H. niger*, which is often known as the Christmas rose. *H. orientalis* was introduced from southeastern Europe in the nineteenth century; growers then began to make selections of superior forms, especially after its gene pool began to be added to by introductions from eastern Europe and the Caucasus.

By the end of the century there were some fifty-odd varieties in Britain and Germany, most of which disappeared over the next century. However, the English nurserywoman Helen Ballard led a revival in the second half of the century, obtaining plants from Germany and even teaching herself German so that she could research the plants. She produced named cultivars from division — an expensive process, since the plants are slow-growing. From the 1980s onwards, however, nurseries began seed production from carefully selected plants, and this has led to greatly increased popularity for the hellebore. The range of colours, including picotees, doubles and spotted forms, is now extensive, and has largely reproduced the diversity the plant had before World War I. The genetic diversity of the plant has also been added to by seed-collecting in the former Yugoslavia in the period shortly before the region descended into civil war in the early 1990s; anyone seeking seed there today must be wary of the landmines with which the combatants liberally sprinkled the countryside.

ORIGIN
H. orientalis, the main source for the garden plant, is native to the Balkans and Turkey

LONGEVITY
Long-lived perennial

SIZE
To about 60 cm (24 in), when in flower

HABITAT
Woodland

COLOURS
Dull reds, greeny whites, dark purples (almost black) and pale yellows

Hyacinthoides non-scripta (Asparagaceae)

English bluebell

Starch for ruffs

Britain has a relatively restricted wild-flower flora and few real spectacles, but when it comes to blue, this bulb is a world-beater. Going to visit the bluebells has been a tradition for many families for more than a century, mainly in their stronghold of southeastern England. Starting just as spring is turning into summer, the flowers can turn woodlands into lakes of blue that in some cases seem to reach as far as the density of tree growth allows us to see.

Like many wild-flower spectacles, there is nothing particularly natural about such a dense growth of one species. Bluebells are the result of woodland management, since they flourish only in relatively light woodland. It is only occasional felling, or the practice of coppicing, that really provides enough light at ground level for the plants to seed and spread enough to provide the density of flower that makes a bluebell wood worth visiting. They are easily suppressed by brambles or bracken, too, and this can happen if light levels are too high for too many years in succession.

Historical records of bluebells are relatively few and far between, and the plant was known to our ancestors more as a source of starch for stiffening the ruffs worn by the wealthy of Elizabethan England than as something to admire. It was not until the nineteenth century that writers took much notice of it, and gardeners too. Small patches of bluebells are rather unimpressive, and it is no surprise that the larger, and paler, Spanish bluebell, *H. hispanica*, makes a better garden plant. Finally, it is worth noting that the common name is an English designation. For the Scots the bluebell is *Campanula rotundifolia*; for the Americans it is *Mertensia virginica*; and Nigerians and New Zealanders have their own, totally unrelated, 'bluebells' too.

ORIGIN
Northwestern Europe

LONGEVITY
Long-lived perennial

SIZE
About 25 cm (10 in)

HABITAT
Light woodland; some exposed grassland areas

COLOURS
Blue with a touch of purple

An immature seed capsule, with the old style still attached.

The characteristic pea family
flower with the large 'standard'
petal and two 'wings' visible.

Lupinus 'Avalune' (Fabaceae)

LUPIN

Spires of many colours

Lupins make an impact. It has something to do with the combination of colour and that powerful upright shape. En masse, and with a range of colours that blend into one another, the effect is powerful indeed. For many visitors to the traditional British flower show, stands with massed ranks of lupins are among the most memorable aspects of the whole experience. The flower spikes are traditionally cut and placed in metal vases, to give an effect that would be impossible from living plants.

Southern Europe has a number of colourful native lupins, one of which, *L. luteus*, is often used as a green manure, turning whole fields golden yellow in the process of adding nitrogen to the soil. European species are all annuals, whereas those that began to be introduced from eastern North America in the seventeenth century were perennial; during the next century more colourful species from the west arrived in Britain thanks to the Scottish botanist and explorer David Douglas.

It was a northern English professional gardener, George Russell, who effectively launched the genus as a garden plant in the early twentieth century. He had been given some seed by an employer and had started making his own crosses, on rented allotments because he did not have space at home. In 1938, at the age of 79, he staged his first exhibit at the prestigious Chelsea Flower Show in London. Over the next two decades he actively pursued his retirement hobby, making the name 'Russell' inseparable from that of 'lupin', his flowers gracing herbaceous borders up and down the British Isles. From the 1960s onwards the plants rather fell from fashion, although some exciting new varieties were bred towards the end of the century, including some dramatic bicolours, something Russell never succeeded in producing.

ORIGIN
North America

LONGEVITY
Short-lived perennial

SIZE
About 1 m (3 ft)

HABITAT
Woodland edge,
open hillside

COLOURS
Blues (although oddly none of the hybrids is as clear a blue as some of the species), yellows, pinky-reds, creams

Matthiola incana (Brassicaceae)

STOCK

Old-fashioned fragrance

Named after the sixteenth-century Italian doctor and naturalist Pietro Andrea Mattioli, this has something of an 'olde worlde' reputation, the kind of flower associated with historic and traditional cottage gardens. The main reason for growing it is the scent, which is fabulous, but never cloying.

As a designation 'stock' is confusing, deriving from a Germanic root word from which 'stick' also comes; it actually covers several different related species, all members of the cabbage family. They were also sometimes referred to as 'stock gillyflowers', the 'gillyflower' being the pink, which has a similar fragrance and was also very popular in the immediate post-medieval period. The single sturdy stem is a character shared with a variety of other coastal plants, including the cabbage itself. As one of the best scents available to gardeners before the great era of nineteenth-century plant introductions, it was widely grown, and our example was regarded as the best species. They were apparently common in gardens during the sixteenth century, in both single and double forms. By the eighteenth century there was a wide range of colours, mostly pastel shades.

Although no rarity, stocks no longer have anything like the ubiquity or status they once had. Grown largely by amateurs, and to some extent by the cut-flower industry, they do not bloom for long enough to make them worthwhile as summer bedding. Today's plants have not been developed much beyond the 'Brompton' and 'Ten Week', names originally applied to strains produced at a nursery at Brompton, London, in the 1750s.

The flower was popular across Europe, and reached perhaps its apogee in the nineteenth century with the weavers of Upper Saxony in Germany. In order to minimize cross-breeding, they made arrangements that each village would produce seed of only one flower colour.

ORIGIN
Central Mediterranean region

LONGEVITY
Biennial or short-lived perennial

SIZE
Potentially to 80 cm (32 in)

HABITAT
Coasts; sometimes naturalized on waste ground inland

COLOURS
Various shades from deep pink to pure white

The seed pod, seen here in its mature state, contains shiny, relatively large seeds that are easy to handle.

A young plant with the characteristic fleshy roots that enable it to attach itself to tree bark.

Phalaenopsis 'Dubrovnik' (Orchidaceae)

Moth orchid

Industrial production

From being playthings and status symbols of the rich to windowsill plants for anybody who can afford a window, orchids have made an incredible transition. At least, some of them have, most notably *Phalaenopsis*. Now for sale across the developed world, *Phalaenopsis* are plants whose requirements of temperature and humidity are remarkably similar to those of human beings, and so are ideal for the conditions provided by our houses. In particular they need the indirect sunlight that is typical of our homes, making it possible to grow them in a variety of spots. A huge amount of high-tech propagation and breeding has further 'domesticated' these plants, making them even more tractable and, crucially, immensely long-flowering. Laboratory techniques have also made it possible to produce them in huge quantities.

Orchids hybridize easily, and it was a *Phalaenopsis* that was one of the first, produced by Veitch Nurseries in Exeter, southwestern England, in 1875. The plants were a minority interest compared to other orchids (see *Cymbidium* and *Laelia*) until the late twentieth century, when their sheer convenience enabled them to break into a mass market. Much has been achieved since that first hybrid, with fuller flowers in a variety of colours, including decorative spotting and blotching. Modern hybrids are tetraploids, which means that they have a double set of chromosomes, manifesting themselves in bulkier plants, including thicker and more durable petals.

Phalaenopsis orchids have become one of the most traded of all plants, with a series of complex exchanges. Varieties may be bred in the United States, tissue-cultured in Japan, grown on in China or Taiwan and then grown to retail size in the Netherlands. Low-cost Asian labour has combined with high-tech breeding and propagation in an oddly close parallel to the electronics industry that now dominates our world.

ORIGIN
South East Asia, down to Queensland, Australia

LONGEVITY
Long-lived perennial

SIZE
To about 70 cm (28 in), when in flower

HABITAT
An epiphyte (i.e. growing on the trunks and branches of trees) in humid tropical forests

COLOURS
Pink through to white, with some pale yellows and much complex patterning

Primula Gold-Laced Group (Primulaceae)

GOLD-LACED POLYANTHUS

British favourite

These are mysterious flowers, so dark brown as to be almost black, or sometimes dark red, with a delicate gold or silver rim to each petal. Their origin is obscure, but they were the first of this now extremely common genetic complex to be grown. 'Complex' is probably the right word to use, since the common polyanthus is, like wheat, a coming together of three species (and, in the case of modern varieties, often more): *P. vulgaris* (primrose), *P. veris* (cowslip) and *P. elatior* (oxlip). All three throw up odd colour forms, and all readily hybridize naturally, too. The English botanist John Parkinson wrote about a red-flowered 'primrose' from Turkey that had been imported in about 1640. It is thought that the gold-laced varieties appeared fairly soon afterwards. They were certainly very popular with the 'florists': plant enthusiasts, usually artisans or craftsmen, who grew particular flowers to exhibit at competitive shows. Hundreds of varieties were bred, distinguished by minute differences, and because they were easy to grow they became one of the most widely grown of all garden plants in Britain in the late eighteenth and early nineteenth centuries.

The arrival of more exotic flowers during the nineteenth century resulted in the almost total disappearance of the gold-laced polyanthus varieties. The polyanthus itself, however, went off in rather more colourful directions, and during the twentieth century it became one of the most common winter flowers, sold on petrol-station forecourts as well as in flower shops and garden centres; the vast majority of the plants produced are kept for only a few months and discarded after flowering. The range of colours is wide, from rich blues through to yellows and pinks – indeed, about as wide a colour range as it is possible to get from one plant gene pool. The flowers also seem to grow bigger every year. The old gold-laced and a number of other more subtle forms are, however, still popular with enthusiasts, their ease of growth from seed their saving grace.

ORIGIN
Parent species: central and northern Europe and western Asia

LONGEVITY
Short- or medium-lived perennial

SIZE
20 cm (8 in)

HABITAT
Woodland and other shaded habitats

COLOURS
Various dark shades with yellow or white edges

A cross section through a 'pin-eyed' flower, where the female stigma is uppermost.

One of the almost spherical seed capsules.

Primula x *pubescens* (Primulaceae)

AURICULA

Grown for display

Like the previous flower, this has deep historical roots as another florist's plant. Derived from two high alpine species, *P. auricula* and *P. hirsuta*, the auricula is first recorded in the late sixteenth century, with Huguenot refugees bringing them to England at about this time. Hundreds of varieties had been developed in Britain and Germany by the late eighteenth century. There was a decline during the nineteenth century, but the auricula's popularity was revived in the late twentieth. Traditionally growers have gathered in auricula societies to organize competitive shows, with plants exhibited by category, each one with a strict set of defining rules that encourages growers to aim at a perfect combination of shape and colour. Today's growers are concerned not so much with competition as with building up collections and growing plants really well. Traditionally, auricula-growers were nearly all middle-aged working-class men; today's are a far more mixed crowd. Not only have auriculas experienced a great revival, but so has the 'auricula theatre', an outdoor shelving unit originally developed to show off and protect the plants when in flower.

Auriculas are really quite extraordinary, so perfectly circular, and with such rich colours, that they can appear artificial. The leaves, and sometimes the flowers, are often covered in a powdery white substance that is dubbed 'farina' (a word that actually means 'flour' in some Latin languages). Some flowers combine green and brown shades, which when added to the farina produce a very unusual appearance indeed. Unlike the closely related polyanthus (see page 130), auriculas are not amenable to mass production. All are relatively slow-growing, and the flowers of many can be easily damaged by rain; consequently, these varieties are grown under cover, and are propagated frequently to encourage flowering. Some are surprisingly old, which says as much about the determination of the growers as it does about the longevity of the plants.

ORIGIN
Parent species: mountainous regions of central Europe

LONGEVITY
Long-lived perennial

SIZE
20 cm (8 in)

HABITAT
Rockfaces, high alpine meadows

COLOURS
Reds, pinks, purples, yellows, greens, browns

Syringa vulgaris 'Madame Lemoine' (Oleaceae)

LILAC

Out in the cold

Lilacs have become one of the most widely distributed garden and landscape shrubs, especially in continental areas with cold climates: Canada, Russia and Ukraine are all great places to see them. They have long been a favourite subject for artists in Russia, too. Relatively easy to propagate, and very resilient, they also make surprisingly good hedges, and parts of southern Sweden in early summer can be unexpectedly exuberant as all the farm hedges burst into a full flowering of soft purple, pink and white.

S. vulgaris is known to have reached central Europe from the Balkans during the sixteenth century, and several other species arrived during the seventeenth, including *S.* x *persica*, which had been grown in Persian gardens for centuries, as well as in the many countries influenced by that country's civilization. A few white and pink forms of the latter appeared over the next few centuries, and it is thought to have become quite widespread by the end of the eighteenth century across northern Europe. The late nineteenth century brought many introductions from China.

Lilac-breeding took a great leap forwards in late nineteenth-century France under the aegis of Victor Lemoine, almost certainly the most prolific and inventive plant-breeder in history. His garden and nursery at Nancy in the east of the country churned out hybrids from a vast array of genera, but it was his lilacs that were definitely his favourites and among his greatest commercial successes. (In fact, it was his wife – who had better eyesight – who actually stood atop the ladder to do the painstaking business of transferring pollen from one flower to another.) Lilacs were popular for 'forcing', growing in greenhouses for early flower, which is rarely done today. Many Lemoine varieties are popular, however; among modern varieties, those bred in Canada are especially important because they can be grown in regions with very cold winters.

ORIGIN
Parent species: eastern Europe, Turkey, Iran

LONGEVITY
Long-lived shrub

SIZE
To 3 m (10 ft)

HABITAT
Woodland edge, scrub

COLOURS
Various shades of purple and pink, creams and pure white; no true blues, reds or yellows

The flowers of double lilacs are
a complex bundle of petals.

A tulip bulb with shoot emerging.

Tulipa 'Queen of Night' (Liliaceae)

TULIP

Driven to madness

This has been one of the most fashionable tulip cultivars for some time now, a fact that may come as a surprise given its brightly coloured fellows. Dating from the 1940s, it illustrates one extreme of the remarkable range of colours shown by modern tulips. Tulips, which are descended from many different species native to Turkey and Central Asia, also come in many shapes and sizes, and today's range is the result of a process of selection and breeding that started in tenth-century Persia and was continued during the Ottoman Empire and then, from the sixteenth century, in Europe.

Tulips are closely linked to the Netherlands, partly because that nation has led the world in both breeding and production for centuries, but also because of the remarkable period known as 'Tulipmania' during the 1630s. Trade was making the country rich, but with little land, spare cash went in novel directions, such as art and… tulips. Speculative investments including sales of 'futures' drove prices higher and higher, with a record of a 'Semper Augustus' bulb selling for 6,000 florins, when a pig would fetch only 30. The collapse of the bubble left many ruined.

A century later the Ottomans imported quantities of tulip bulbs from the Dutch, setting off a frenzy of breeding among the merchant class and at the sultan's court. Sultan Ahmed III became obsessed by them, and held extravagant tulip festivals, which contributed to a financial crisis and a rebellion by his soldiers, who then forced his abdication.

Returning to sanity, Dutch breeders continued to work on the flowers. During the nineteenth century they worked with many new species, often introduced by Russian botanists, to produce today's range of garden types.

ORIGIN
Parent species: southern Europe, Turkey, Iran, Central Asia

LONGEVITY
Short-lived bulb

SIZE
To 40 cm (16 in); wild species are usually shorter

HABITAT
Open habitats, such as steppe grassland

COLOURS
Reds and pinks through to pure white and then through oranges to yellows; a wide range of purples; no blues

Viola x *wittrockiana* (Violaceae)

PANSY

Darwin's darlings

Violas, of which the group of large-flowered varieties is known collectively as pansies, have a long history of very varied symbolism. Medieval Christianity associated them with death and rebirth, possibly a reference to the ability of the wild species, *V. tricolor*, to appear suddenly as a field weed at the end of winter. The name 'pansy' comes from the French *pensées*, thoughts and memories. Other languages associate the flowers with love, as with the Portuguese name *amor-perfeito*, perfect love. Folk cultures add a more sexual edge, as with the traditional English 'Jack-jump-up-and-kiss-me'.

The development of pansies, as opposed to violets (see page 89) began during the early nineteenth century with the artist Lady Mary Elizabeth Bennet, who collected every variant of *V. tricolor* she could find in her garden in England's Thames valley. Other wealthy gardeners in the area began to collect and make selections, too, and their head gardeners almost certainly would have made connections between them. Sarah Siddons, a well-known actress, fell in love with the flower and brought it to popular attention. By mid-century it had become a 'florist's flower', with artisan growers forming societies and holding competitions; Charles Darwin noted 400 named cultivars in 1835, which he found both attractive and interesting. Such varieties were rather highly strung, and it was not until an English nurseryman who worked at Versailles, John Salter, made crosses with other *Viola* species that there were plants robust enough for winter bedding out and hence very early flowering in the open garden.

With their flowers that look as much like faces as any, pansies have never lost the widespread popularity they had gained by the late nineteenth century. Investment in the breeding of new varieties remains high for markets that include a great deal of public as well as private planting.

ORIGIN
Parent species: northern and central Europe, Russia

LONGEVITY
Annual or very short-lived perennial

SIZE
To 25 cm (10 in)

HABITAT
Waste ground, fields, woodland edge

COLOURS
Predominantly purples and yellows, often in combination

*A seed capsule that has just split open,
revealing the soon-to-be-scattered seed.*

The papery bract that surrounds
the flower head has just
opened to reveal the buds.

Agapanthus campanulatus (Amaryllidaceae)

AFRICAN LILY

Tougher than it looks

Appearing along roadsides, in parks (sometimes under trees) and outside office buildings across warm temperate and subtropical climates, the blue of agapanthus is now ubiquitous. Occasionally a white form is seen, but almost never the rich range of blues and purples that the plants' many hybrids show; anything other than the basic species is unfortunately a rarity. It took a long time for botanists to decide what this plant –first discovered by Europeans in the early seventeenth century – actually was. The great English botanist John Parkinson thought it was a narcissus, and since then it has been placed in five different botanical families. DNA evidence now does in fact put it as a distant *Narcissus* relative.

In Victorian times and well into the twentieth century British gardeners typically grew agapanthus in half barrels; they did not need repotting often, and indeed seem to flower well when growing closely packed. The barrels would be lifted, or dragged, into greenhouses and cold frames to survive the winter. In the early twentieth century some more adventurous growers tried them outside, and it is now accepted that they are indeed hardy in many more areas than was originally thought. European colonists took them around the world, and the plants proved invaluable. Although slow to get going, once established they are more or less indestructible; in particular, their dense mat of steadily expanding rhizomes prevents weeds from growing, making them ideal for low-maintenance situations. This has unfortunately meant that they have become invasive in some places, smothering local flora, especially in Australia. However, a breeding programme is now producing sterile ones that won't seed into the wild.

The ubiquity of blue agapanthus in many countries is in a way a pity, since there are so many hundreds of varieties to choose from, available in every shade from deepest purple to pure white. There are also a growing number of dwarf ones. Many of these are grown in Britain's national collection of the plants, which – curiously – is in the northern county of Yorkshire, not the balmy coastal location one might expect.

ORIGIN
South Africa

LONGEVITY
Long-lived spreading perennial

SIZE
To 1.5 m (5 ft), when in flower

HABITAT
Open situations, often montane

COLOURS
Many shades between purple and blue, plus white garden forms

Anemone x *hybrida* 'Honorine Jobert' (Ranunculaceae)

Japanese anemone

Born to survive

These plants, with their masses of bowl-shaped flowers that persist over months, are survivors and can mark out old gardens more effectively than almost anything else. They are tall, robust and – eventually – strongly spreading, although typically slow to establish; many are the gardeners who have planted one and then, seeing that it was exactly the same size three years later, decided it was 'unhappy' and tried to move it, discovering that it had not been wasting its time, but had been growing a massive root system. Cutting into these roots stimulates the anemone into producing more plants.

The plant is actually of Chinese origin, but had been grown in Japan for centuries, never attracting much attention from growers. It was the Scotsman Robert Fortune who first introduced it to Europe, in 1844. Fortune was one of the most successful of the early plant-hunters, particularly since he was able to explore parts of the country in disguise, as the Chinese government at the time forbade foreigners from leaving the tiny areas of treaty ports. His main task was smuggling out tea plants for the British to grow on in India.

These anemones soon became staples of the herbaceous perennial market, and they cover much ground in the gardens of grand old houses, often thriving under trees. Most garden varieties are in shades of pink, although our 'Honorine Jobert' is pure white; it was a remarkably early mutation (in 1851) and is still probably the finest and most robust variety. Recent years have brought the introduction of relatively dwarf varieties, bred by nurseries in the United States.

These plants have recently been reclassified as *Eriocapitella*, a designation that was originally suggested for a division of *Anemone* by the Japanese botanist Takenoshin Nakai in 1941. DNA analysis has now lifted this name to a higher status, that of being an independent genus.

ORIGIN
Far East, from eastern Siberia down to southeastern China

LONGEVITY
Long-lived spreading perennial

SIZE
To 1.5 m (5 ft), when in flower

HABITAT
Woodland edges and scrub

COLOURS
Many shades of pink, plus white

The seeds are contained within a cotton
wool-like mass of white fibres.

Seeds are attached to long,
feathery structures, which here
have not quite matured.

Clematis x *jackmanii* (Ranunculaceae)

CLEMATIS

Clambering for attention

Clematis are now one of the most important groups of garden plants, with dwarf ones, ideal for small gardens, balconies and even window boxes, selling in their millions. The plants have, however, come a long way. The very modestly flowering European species appear to have been grown in gardens from the sixteenth century onwards, but it was the opening up of China and Japan in the nineteenth century that led to the large-flowered hybrids we know today. Far Eastern growers had for centuries had plants with showy flowers and, crucially, a tendency to flower on side shoots. This ability to flower low down makes them very useful as garden plants, as is shown by the habit of growing them on obelisks made from wooden trellis.

A breakthrough was made in 1858 by the English nurseryman George Jackman, who crossed an existing hybrid with the European *C. viticella* and the East Asian *C. lanuginosa*. The resulting showy, vigorous plant proved a huge success. Meanwhile, *C. montana* had arrived from the Himalayas, introduced by the wife of the governor general of British India. It too was a great success, clambering up the sides of British country houses, along garden walls and even to the tops of quite substantial trees, smothering everything with pink flowers for a few weeks in early summer. From the great botanic gardens of St Petersburg came *C. tangutica* in the late nineteenth century, a botanical outcome of the 'great game', when British and Russian explorers were both investigating, and seeking to dominate, Central Asia. It and similar species are vigorous, and their strangely thick yellow petals are borne, usefully, in late summer.

The Swedish nurseryman Magnus Johnson is generally thought of as the greatest clematis-breeder of the twentieth century. Over the waters of the Baltic, however, good work was being done in Poland and the Soviet republics of Estonia and Latvia, and one of the finest varieties of recent years, *C. viticella* 'Polish Spirit', was bred in the mid-1980s by a Polish monk, Brother Stefan Franczak, in Warsaw.

ORIGIN
Globally distributed across most cool temperate zones

LONGEVITY
Long-lived climber

SIZE
To 3 m (10 ft) (this variety); others between 30 cm (12 in) and 10 m (33 ft)

HABITAT
Woodland edges, scrub

COLOURS
Deep purple-blue (this variety); species include blues, purples, pinks, reds and yellows

Cosmos bipinnatus 'Sensation' (Asteraceae)

COSMOS

Bright stars

The botanical name of these bright and cheerful daisies comes from a Greek word that covers both 'ornament' and 'order', hence its use also to describe the heavens and the arrangement of the stars. The name was chosen by Antonio José Cavanilles, director of the Royal Botanic Garden of Madrid, which was the point of introduction for the plant. From there it was distributed across Europe, with forms being selected for colour (deep pink through to white) and increasingly for different heights and various petal shapes.

Garden cosmos are annuals but are frost-tender, so are generally planted out in early summer from young plants raised under cover. However, in slightly warmer climates, or in continental ones that experience a rapid rise in spring temperatures, they can be sown into the ground without fear to germinate and grow. As a result they have become one of the most commonly grown warmer-climate annuals. They produce plentiful seed, so it is cheap, which has perhaps led to their overuse in highway beautification schemes. From the 1980s onwards wild flowers were increasingly used along roads in the United States, with Lady Bird Johnson (the former First Lady) heading a campaign from her home in Texas. In the early years there was a tendency to use annuals for a quick effect, rather than locally native and more durable perennials, and cosmos was one of the commonest, so for many Americans the flower is strongly associated with roadsides. A vigorous seeder, it establishes easily outside gardens and has become something of a weed in South Africa and Australia.

C. atrosanguineus is a somewhat mysterious relative, a short-lived perennial with dark red flowers and a distinct smell of chocolate. It became popular in the 1990s, and it was widely believed that the plant was extinct in the wild. In the 2010s, however, Mexican botanists found the plant in three counties in that country, and collected seed – a salutary reminder of all the natural world can hold even when we are unaware of it.

ORIGIN
Southern and western North America, down to southern Brazil

LONGEVITY
Annual or short-lived perennial

SIZE
Under 1 m (3 ft)

HABITAT
Seasonally dry open situations, woodland edges

COLOURS
Pinks and yellows, some reds

The seeds of cosmos just before dispersal. The bracts at the bottom have protected the developing flower head.

*The column, the core of the flower,
is a unique structure containing
both female and male organs.*

Cymbidium 'Rembrandt Masterpiece' (Orchidaceae)

CYMBIDIUM ORCHID

Oriental prize

Cymbidium flowers are never really showy. Their colours are always oddly muted, and much of their beauty lies in their delicate, intricate markings, often in greens and browns. That the flowers are very long-lasting (months rather than weeks) has been one reason for their popularity, although the tough quality of the blooms has led some to liken them to plastic. This toughness makes them good buttonhole or corsage flowers.

Traditionally, orchids were regarded as 'difficult'. The cymbidium, however, is one of those that in the right location seems almost indestructible. In Mediterranean climates, for example, old plants in pots seem to thrive on balconies, steps and courtyards – anywhere that is safe from overnight frost and out of full sun.

Cymbidiums can be found wild in mountainous areas of southern China and parts of Japan, and they have long been cultivated there. They were first recorded as being grown in China around 200 BC, and were apparently mentioned by Confucius in around 500 BC. They were taken to Japan during the Heian period by Buddhist monks, who then started making selections from wild plants in their own country. Along with many other flowers, they became very popular in the Edo period, being grown by a wide range of people, although, oddly, the most prestigious and expensive ones were those selected for having variegated leaves rather than distinctive flowers.

The very wide range of modern cymbidium hybrids is derived from a gene pool of about ten species, most breeding having happened in the latter part of the twentieth century in those climates where they can be grown outside, such as California, Australia and New Zealand. The hybrids are much bigger and burlier than the delicate wild species, and are tough, resilient plants, while the fact that the various wild species flower at different times has ensured that modern varieties are available to flower across a range of seasons. A dwarf species, *C. pumilum*, was used to produce a range of miniatures from the 1970s onwards, but those have never really caught on.

ORIGIN
Mountainous areas of southeastern Asia, warmer parts of the Far East, Asian archipelagos

LONGEVITY
Long-lived perennial

SIZE
Cultivated varieties up to 1 m (3 ft); wild species generally smaller

HABITAT
Seasonally dry open woodland, rocky ground

COLOURS
Pinks, yellows, browns, greens, creams

Freesia x *kewensis* (Iridaceae)

FREESIA

Softly scented

Freesias are well known but at the same time low-key, often components of bouquets of larger but more spectacular flowers, as members of the supporting cast rather than stars. That they are important for the global floristry industry says as much about their ease and speed of growth as it does about the qualities of the flowers. The flowers, though, have it all, being reasonably long-lasting in the vase, available in many colours and delightfully, but never overpoweringly, scented.

The story of the freesia is itself quite low-key. There are about 12 species of these South African bulbs, discovered by Europeans in the late eighteenth century, but – unlike with many other South African plants – there was no significant horticultural interest in them until the late nineteenth century. The German botanist Max Leichtlin found some yellow-flowered plants of what is now called *F. leichtlinii* in the Botanic Garden at Padua, Italy, in 1874; he distributed the seed and the plant became quite popular as a greenhouse plant, for its early fragrant flowers. *F. alba* first appeared in the English nursery trade in 1878, and spread quickly to the rest of Europe and North America. It was clearly something of a sensation, being written up in a great many garden journals of the day. Soon afterwards hybrids were made, with several other species being added to the mix. The vast majority of freesias are grown in greenhouses and polytunnels by professional growers for the cut-flower trade, rather than as garden plants; indeed, they are hardy only in frost-free climates.

The botanical history of the freesia is a complex one. It is one of those genera that botanists have never seemed to agree on, with species being named a *Freesia* and then being evicted from the genus, while others start off as something else before being merged into *Freesia*. The truth is that there is a plethora of small South African bulbs, and it is largely chance that freesias have become as popular as they have, rather than *Babiana, Sparaxis, Tritonia, Moraea* or *Ixia*.

ORIGIN
South African, mostly
Cape Province

LONGEVITY
Long-lived perennial

SIZE
About 60 cm (24 in)

HABITAT
Light shade in scrub
and woodland edges

COLOURS
Pinks, yellows, reds,
creams, pinks, purples

*A fibrous coat conceals a swollen
stem that acts as a storage
organ, making this technically
a 'corm' rather than a 'bulb'.*

*Fluffy gazania seed has evolved
to fly away on the breeze and
distribute the species.*

Gazania 'Kiss Bronze' (Asteraceae)

GAZANIA

South African beauty

Gazanias' ability to cover the ground with dense foliage and send forth a long season of short-stemmed, brightly coloured daisies is greatly valued by domestic gardeners and landscape architects in warmer climates. As well as covering banks in dry climates, they have a particular usefulness by the coast, being tolerant of wind and salt. This paragon of the plant world is named after a fifteenth-century Greek scholar: Theodorus Gaza. The man himself was of great importance to learning, and translated into Latin an important early work of botany (that of the Greek Theophrastus), but he was not noted as a botanist, so the fact that he had a plant named after him illustrates a fundamental absurdity of the whole system of botanical naming. Never mind; he is worth remembering.

Gazanias are among the plants that give the drier inland areas of South Africa their incredible reputation for brilliant wild-flower colour, at least during the rainy season. Their habit of growing flat against the ground and spreading slowly is very distinctive, and was spotted by both South African growers and nurserymen elsewhere, who eagerly imported seeds. Various wild species were crossed during the latter part of the twentieth century, and selections made, and the result is a range of strongly coloured plants that can be grown either as perennials or, in colder climates, as annuals, since they flower within a year from seed.

Gazania is one genus in the daisy family among many from South Africa, which has experienced an explosion of genetic diversity unrivalled anywhere in the world. Along with *Osteospermum* and *Dimorphotheca* (whose names are a good deal less easy to remember), these easy plants have transformed gardens and public spaces the world over. They may be simple to propagate and spread readily, but none has yet become an invasive problem.

ORIGIN
South Africa

LONGEVITY
Long-lived perennial; often grown as annual in regions with cold winters

SIZE
To 30 cm (12 in)

HABITAT
Open habitats, coastal dunes, pastureland

COLOURS
Fiery: oranges, reds, yellows

Hydrangea macrophylla 'Bodensee' (Hydrangeaceae)

HYDRANGEA

Where it rains

Lining many a driveway, these are among the most successful shrubs in cultivation; where hydrangeas grow well, they can become ubiquitous. And where they don't, the fact that they are so easy to propagate from cuttings, and so quick-growing, means that they are often grown as disposable pot plants. Among the many plants popular with growers in eighteenth-century Japan, they attracted great interest from early Western travellers to the country. There was a problem, however, because many of these early introductions were 'mopheads', where the sterile flowers that normally just ring the fertile ones have taken over the entire flower head. Botanists did not know how to classify them until eventually it was realized that they were closely related to plants that had been introduced from North America somewhat earlier – they were all *Hydrangea*.

Mophead hydrangeas were all the rage in mid-nineteenth-century Europe, where growers took advantage of the fact that a plant of flowering size can be grown from a cutting in the space of a year. Millions were sold, but comparatively few planted out in gardens. Their love of mild winters, acid soil and constant moisture meant that their use tended to be restricted to particular climate zones: the wetter the better. In places with these temperate climates, such as the Azores, northern Portugal and Ireland, they have become major features of the landscape.

Hydrangea is a varied genus, and one where hybridization seems easy. The number of crosses leading to successful varieties over the last century and a half has been in the thousands. Many of the more recent ones are more tolerant of cold or less dependent on moisture, so new varieties have made inroads into territories that were formerly considered unsuitable for them. The ivory-white 'Annabelle' has been particularly successful; hardier, smaller, more delicate and less shrubby than the conventional hydrangea, this plant has taken over so many gardens in Holland that it is arguably now more typical of the country than the tulip.

ORIGIN
North America, Far East

LONGEVITY
Long-lived shrub;
some climbers

SIZE
2–3 m (6–10 ft)

HABITAT
Woodland, woodland
edge, scrub

COLOURS
Pinks and, depending
on soil chemistry, blues;
many creams and whites

A single flower from a hydrangea flower head; the vast majority of these are sterile.

*The withered remnants of the flower
are still attached to the
dried seed capsule.*

Ipomoea tricolor 'Heavenly Blue' (Convolvulaceae)

MORNING GLORY

One-day wonder

These annual climbers, usually grown in warmer climates, add colour to whatever they are sent to climb up. They are particularly good at covering – and hiding – old fences or anything else unattractive, at least for a few months of the year. Our variety is a lovely clear blue, but they come in many other colours, and this variability is part of what makes them attractive.

Introduced from China to Japan some time before 1000 AD, the morning glory became a popular garden plant there among all social classes, and a very popular subject for artwork over many centuries. The fact that the flowers live for only a day, and that the plants themselves are only annuals, appealed to the Japanese love of transient beauty. Climbers were also appreciated, since they took up little space and could make useful screening during hot summers. During the Edo period the selection of new varieties became very advanced, with growers selecting not just for conventional beauty but also for *demono* plants with bizarre leaf and flower shapes – not plants that could in any way be described as beautiful. Surviving correspondence between growers of these plants reveals that they understood the principles of what we now know as Mendelian genetics – the word *demono* actually means 'recessive', as in recessive genes. As with several other plants (among them camellias and irises), one of the schools of morning-glory cultivation was practised in the region dominated by the Higo clan, around the southern city of Kumamoto.

Another plant that is often dubbed 'morning glory' is the long-lived perennial relative *I. indica*, which can grow to at least 10 m (33 ft) and smother banks, trees and even buildings. The sweet potato, *I. batatas*, is also a member of the genus. The seed of all species contains potent alkaloids, including compounds similar to LSD, leading to their being used both in herbal medicine and as psychoactive drugs for use in religious ceremonies among Indigenous communities in South America.

ORIGIN
Far East

LONGEVITY
Annual climbers

SIZE
2–3 m (6–10 ft)

HABITAT
Forest margins, waste ground

COLOURS
Pinks, red-pinks, blues, whites

Lonicera periclymenum (Caprifoliaceae)

HONEYSUCKLE

Scent of the hedgerows

One of the most delightfully scented of all European plants, the familiar garden honeysuckle is very similar to the wild plant, which flourishes along hedges and at the edges of woodland but can also be found as a ground-level scrambler, especially on poorer acid soils. Its old name in English is 'woodbine', which refers to its twisting habit and the fact that the stems were often used as a twine by country people. Somewhat ironically, 'woodbine' was used as a brand name for a popular British brand of cigarette, widely smoked throughout the earlier part of the twentieth century and very much a marker of working-class life.

Honeysuckle's sweet smell and its twining nature has long given it an association with romantic love in many European cultures, as has its tendency to be found growing along the kind of path that lovers would have chosen in country districts. The plant was popular in the seventeenth century but often not as a climber, instead trained up a pole and then kept as a standard. Several forms were introduced from Holland and Flanders, and these are still popular today. They do not make an impact in the way that other climbers (roses, clematis, wisteria) do, but integrate nicely with them, thrusting their fragrant flowers out from among the growth of others.

American honeysuckles are mostly bright orange, but scentless – both signs of having evolved for pollination by hummingbirds, as opposed to the pale but fragrant European species, whose target is moths. The Americans were first introduced to Europe in the seventeenth century, while several Asian species arrived during the nineteenth. Indeed, the centre of diversity for the genus is China, which has 100 out of the 120 species. One Asian species, *L. japonica*, has in parts of the United States become a hated invasive alien that smothers native vegetation.

ORIGIN
Eurasia; other species in North America

LONGEVITY
This species: long-lived climber; many other *Lonicera* are free-standing shrubs

SIZE
This species: to 3 m (10 ft); others can be considerably larger

HABITAT
Woodland edges, moorlands

COLOURS
Pink, creamy white, almost yellow

The characteristic tight cluster
of honeysuckle berries.

The seed head with its hundreds
of small seeds ready to be
dispersed (or eaten by birds).

Rudbeckia fulgida (Asteraceae)

BLACK-EYED SUSAN

Washington daisies

The bold yellow flowers with dark centres make rudbeckias popular plants for parks and gardens. Unlike most late-flowering plants in the daisy family, many are usefully short, especially *R. hirta* and *R. fulgida*. Both arrived in Europe in the early eighteenth century, but did not get much attention from gardeners until the later years of the nineteenth. Superficially very similar, they have different life cycles. *R. hirta* is fast-growing, usually flowering in its first year, but rarely lasts more than a year or two. The plants have remained popular in summer bedding schemes for well over a century now, receiving a boost from breeders in the 1950s with a notable variety, introduced as 'The Gloriosa Daisy', with extra-large flowers.

R. fulgida is soundly perennial, and spreads steadily. It was not grown widely until the middle of the twentieth century, but made a breakthrough with a form of *R. fulgida* var. *sullivantii* that was spotted in the Botanic Garden of Graz, Austria, by a colleague of the great German nurseryman and plant-breeder Karl Foerster. Having been given some of these plants, Foerster named the form 'Goldsturm' and began propagating it during World War II. Despite being just outside Berlin, and someone of whom the Nazi regime was suspicious (his brother had already had to flee to Switzerland), Foerster and his nursery survived, and by 1945 he had about 5,000 plants. It is interesting to note that while British nurseries were legally obliged to turn their land over to food production during the war, this did not happen in Germany until much later in the conflict.

Rebuilding Germany in the 1950s involved reconstructing cities, housing and public parks, and Foerster's students were in the vanguard of designing and planting new landscapes. 'Goldsturm' (along with other rudbeckias) proved ideal. It did to Wolfgang Oehme, too, a German gardener who emigrated to the United States and went on to introduce a distinctive style of perennial planting there, often featuring vast swathes of 'Goldsturm', starting with some key areas of Washington, DC.

ORIGIN
North America, mostly east of the Rockies

LONGEVITY
Perennial; a few short-lived perennials

SIZE
This species: to 1 m (3 ft); others mostly taller

HABITAT
Forest margins, savannah, prairie

COLOUR
Yellow with a dark centre

Solidago rugosa (Asteraceae)

ROUGH GOLDENROD

Rubber substitute

Goldenrods – for many the quintessential autumn flower – were one among the many daisy-family plants introduced to Europe in the late eighteenth and early nineteenth centuries. By the end of the latter century, various forms of two particularly vigorous species, *S. gigantea* and *S. canadensis*, were thriving as late-season border plants; they were popular as cut flowers, too. Both species are very vigorous, and by the middle of the next century they had well and truly escaped. They began to cover areas of waste ground in Germany, while in Britain it was the suburban rail network that took them far and wide, helped by domestic gardeners throwing unwanted plants over the back fences of gardens that lined the railway cuttings.

By the late twentieth century the plants were thoroughly unpopular, and had given the whole genus a bad reputation. That is a shame, because many are garden-worthy plants, such as our example, *S. rugosa*, which spreads only slowly, looks very elegant and attracts butterflies in hordes. It is one of the plants that the pioneering American landscape architects Wolfgang Oehme and James van Sweden helped to popularize. There are many other species for a variety of habitats, and since all are good pollinator plants, their rehabilitation – particularly in the context of the current interest in native plants and ecological planting – seems well on the way and much deserved.

The strangest part of the goldenrod story concerns the American inventor Thomas Edison and his discovery that goldenrod contains enough latex to be a usable substitute for imported rubber from tropical rubber trees – especially if your country is at war and rubber difficult to import. From 1915 onwards he extensively researched the possibility of United States-produced goldenrod rubber. By the time the next war came around, however, there was synthetic rubber, and Henry Ford's decision to use that instead spelled the end of Edison's project.

ORIGIN
North America, mostly
east of the Rockies

LONGEVITY
Long-lived perennial

SIZE
1–2 m (3–6 ft)

HABITAT
Forest margins,
savannah, prairie;
some wetland species

COLOUR
Yellows

A single goldenrod flower — one of many on the branching stems.

The distinctive, rigid nasturtium seed
capsule contains three large seeds.

Tropaeolum majus (Tropaeolaceae)

NASTURTIUM

Bright and piquant

Known as cheerfully bright flowering plants of summer – albeit with brief lives, since the first frosts will turn them to mush overnight – nasturtiums have a peppery flavour, and so the leaves and flowers have often been used in salads. All are South American plants, variable at first sight, but having a tendency to be climbers. The Andean *T. tuberosum*, known as mashua, is a perennial climber long grown by and indeed still popular with Indigenous peoples for its starchy tuberous roots, a reminder that the familiar potato is simply one among many such cultivated plants from this region.

T. minus appeared in Europe at the end of the sixteenth century via the Spanish Empire, and was well established in popular cultivation 100 years later. However, its bigger and flashier cousin *T. majus* arrived in the late seventeenth century and gradually displaced it. Free with its easily germinated seeds, it became a popular garden flower; the elite, however, could grow doubles, which could be kept going by taking cuttings that had to be overwintered in frost-proof conditions – a much more expensive business. A double that was fertile and so could be grown from seed appeared in a garden in California in the 1930s; sold as 'Gleam', it rapidly became hugely popular.

In many languages the plant is called a cress, because of its flavour. The German version, *Kapuzinerkresse* or Capuchin cress, is named for the shape of the flower, which, with its long nectary, resembles the cowl of this monastic order.

The flowers of *T. speciosum* are much smaller, but again in a vivid, hot colour, with a similar shape. This climbing plant, introduced from Chile to Britain in the mid-nineteenth century, became strongly associated with those gardens where it grows well, generally in mild, wet, westerly locations, where it can climb up an evergreen tree such as a yew and show off its scarlet flowers in late summer.

ORIGIN
Temperate South America

LONGEVITY
This species: annual;
others mostly perennial

SIZE
1–2 m (3–6 ft)

HABITAT
Stream sides,
forest margins

COLOURS
Oranges, yellows

Zantedeschia aethiopica (Araceae)

ARUM LILY, EASTER LILY

Fit for a funeral

In a quiet way, this has been one of the world's most successful plants. From its native South Africa, and an introduction to Europe in the early eighteenth century, it has been spread throughout the globe, largely during the twentieth century. It is a robust, long-lived perennial that just needs a seasonally wet patch to settle down in, after which it will be there for decades. It can spread and be invasive in wetlands, but generally it stays put and becomes a permanent part of the scenery, delivering its blooms every spring with no maintenance whatsoever.

The 'flower' itself is not a flower at all, however, but a large bract enclosing a column that contains a great many flowers, each a few millimetres across, the males above and the females below, pollinated mostly by beetles. As an adapted leaf, rather than a flower, the showy bit of the arum lasts a long time, and so has long been popular with florists, which is one reason for its wide distribution. It is also quite unlike anything else, and so has become an important flower in many cultures; some Indigenous communities in Mexico, for example, make great use of it in their places of worship and in their art.

For some reason, arum lilies became popular for funerals during the late nineteenth century in Britain, and so their use at other times became almost taboo for many. Its flowering at Easter linked it to the Easter Rising of 1916, when guerrilla fighters took over the centre of Dublin in a near-suicidal mission against British rule. It has been central to more extreme Irish Republican iconography ever since, including playing its part in the funerals for members of the Irish Republican Army killed during the Troubles in the six counties of Northern Ireland during the 1980s.

ORIGIN
Eastern South Africa

LONGEVITY
Long-lived perennial

SIZE
1 m (3 ft)

HABITAT
Stream sides, wetland

COLOUR
This species: pure white; others yellow, purple-black, pink

The tightly packed column of fleshy
fruit is typical of the arum family.

A single aster disc floret; it is the outer ray floret that has the showy, coloured petals.

Aster 'Little Carlow' (Asteraceae)

ASTER, MICHAELMAS DAISY
St Michael's daisy

We all love daisies, and this appreciation of open, sunny flowers has led us to cultivate a great many of them, especially since their numbers seem to build in late summer and reach a peak in autumn. Among these, one of the biggest groups are what are dubbed Michaelmas daisies, from the timing of their flowers around that Christian festival at the end of September. In theory, these are derivatives of *A. novi-belgii*, which was introduced to Europe from the New York area in 1687. In practice, the name is applied to many other American species of *Aster*, which were introduced over the next couple of centuries, and particularly loved for this late flowering. They reached a high point in cultivation in the early years of the twentieth century, and British and German nurseries raised a vast number, many of them hybrids between the various species. Catalogues and books of the time illustrate great bunches of the flowers, making the most of the fact that for the first time printing technology was just about advanced enough to show colours in a reasonably true likeness.

Many of these late-flowering daisies were high-maintenance: prone to disease and needing to be dug up and divided every few years. Not surprisingly, interest slowly began to shift to the easier-going and more robust species, and the hybrids between them. Today, it is almost entirely those that we grow. Among them is 'Little Carlow', bred by one Mrs Thornley in the 1930s in Devizes, southwestern England, from two common species from the American northeast. It is still unrivalled today for its reliability in producing generous sprays of blue flowers above slowly expanding clumps of healthy foliage for weeks on end during late summer and early autumn, and for doing so for years without attention.

'Aster' was always a vague designation, however. Botanists have now excelled themselves in making a more accurate definition of the genus, showing the real relationships between species, but seeming to choose new names that are not only unmemorable but also unspellable. Our particular *Aster* is now a *Symphyotrichum*!

ORIGIN
Derived from species of eastern North America

LONGEVITY
Long-lived perennial

SIZE
1 m (3 ft)

HABITAT
Parent species: woodland edges, abandoned fields, roadsides

COLOURS
Blue tinged with purple

Banksia coccinea (Proteaceae)

SCARLET BANKSIA

Childhood favourite

Many people know banksias without having actually met one. Australian children are familiar with the Big Bad Banksia Men, villains of the Gumnut stories by May Gibbs, a leading writer and illustrator of children's books during the late twentieth century, featuring characters based on the wildlife and wild plants of the Australian bush. The hairy, knobbly Banksia Men are based on the other reason many are familiar with banksia products rather than banksias themselves: the distinctive seed pods, rather like pine cones but woodier and more irregular. These are adapted to hold the seeds until a bush fire burns through, liberating them to start a new life. The collecting and turning of banksia cones on a lathe is a minor cottage industry in Australia, and the resulting artefacts make long-lasting souvenirs that have found their way to mantelpieces and shelves all over the world.

B. *coccinea* is from southwestern Australia, where an extraordinary flora reaches a peak of inventive biodiversity. It is the most colourful of some 60-odd species that are found in a relatively small area. All have extravagantly large flowers that can be most realistically likened to plastic brushes, and most are pollinated by birds. The plants themselves are usually shrubs of small to medium height, although some are prostrate ground-cover plants; all have dramatic foliage, often looking as if it were cut with scissors or crimping shears. Easy enough in dry summer climates on poor soils, the plants rapidly die in 'better' conditions. They were mostly described during the nineteenth century, but it was only in the late twentieth that they became at all important as cultivated plants. Our species was first collected by European botanists in 1801, and was named some ten years later. It is grown commercially in Israel and the United States for the floristry industry, the longevity of the spectacular flowers being an important selling point.

ORIGIN
Extreme southwestern tip of Australia

LONGEVITY
Long-lived shrub

SIZE
8 m (26 ft)

HABITAT
Part of a scrub community

COLOUR
Scarlet

A typical banksia seed head, in which, as is typical, only a few of the seeds have reached maturity.

*Male and female flowers are
different in begonias; this
is the smaller female.*

Begonia 'Kimjongilhwa' (Begoniaceae)

BEGONIA

Favoured by a dictator

Brilliant-scarlet blooms present a bold and bombastic face to the world, mirroring the stance taken by the late dictator of North Korea, Kim Jong-il, infamous for his combination of an advanced missile system and frequent famine among his country's population. Produced originally by a Japanese breeder, the flower was chosen as his emblem because it flowers for his birthday in mid-February; it was used to adorn his body as it lay in state after his death in 2011, and still provides floral decoration for state events.

This genus provides many intense colours, mostly at the 'hot' end of the spectrum. It is a very large genus, with some 2,000 species, a reflection of extraordinary genetic flexibility and diversity. Begonias appear to develop new species very rapidly, and, since many come from tropical mountain regions with highly fragmented and diverse habitats, there is plenty of scope for new species to evolve. This genetic flexibility has also led to their becoming very popular garden plants since their introduction to Europe and North America in the nineteenth century.

The vast *Begonia* genus is divided into several distinct subcategories, and of these only some have entered cultivation. Of those, the 'tuberous' category has long been one of the most important, with plants that rapidly produce very large flowers from a tuber that, like that of a dahlia, can be kept from year to year out of the ground. This makes them very useful for summer bedding displays in strongly seasonal climates. These versatile plants are descended from wild species of Andean cloud forests, first bred by the innovative Veitch nursery in England in the 1860s.

Other categories of begonia often combine attractive foliage and flowers, such as the Elatior types, making them commercially important pot and display plants for a longer season. There are others that are grown only for their attractive foliage, such as *B. rex*, a popular houseplant tolerant of relatively low light conditions. Yet other, taller 'cane' types are popular container or garden plants for warm climates.

ORIGIN
Widespread across tropical climates and adjacent slightly cooler climates

LONGEVITY
Long-lived perennial

SIZE
Tuberous varieties rarely exceed 50 cm (20 in)

HABITAT
Mostly montane; usually moist, shaded habitats

COLOURS
'Hot' colours (tuberous varieties); usually pink or white (others)

Bougainvillea glabra (Nyctaginaceae)

BOUGAINVILLEA

Coastal colour

The quintessential warm-climate climbing plant, bougainvillea clambers and sprawls over countless houses, balconies and walls in more or less any part of the world with a minimum temperature no lower than just below freezing. Its purple is outrageous and long-lasting, the reason for the latter quality being that the purple is in fact a mass not of flowers but of bracts – leaves that attract pollinators to the very small white flowers at their centre. The bracts are there before the flowers open and for a long time after they die. It is not clear what evolutionary advantage accrues to the plant from this, unless perhaps we argue that it was realized in the Age of Empire, when Europeans came along and fell in love with them, enabling the plant's spread all over the globe. The first Europeans to set eyes on bougainvillea were the French botanist Philibert Commerson and his assistant and partner Jeanne Baret, in about 1780. Baret was in fact disguised as a man, but it may be that she was the first woman to circumnavigate the globe.

Whether there are three species of bougainvillea or as many as 18 is a matter of continued dispute among botanists. However, the plants interbreed easily, and hybrids are found in the wild and very frequently in cultivation, hence the 300-odd varieties that are recognized today. The plants also produce 'somatic mutations', when a stem suddenly changes character (apple trees do the same); cuttings taken from this preserve the new character, so creating yet another new variety. The results are plants with red, pink, cream or yellow-orange bracts, and varieties that are more or less vigorous, including some dwarfs.

The success of bougainvillea has been helped by it being easy to grow, suffering few problems with pests or disease and being notably tolerant to salty air, making it an ideal coastal plant. It is immensely long-lived and vigorous, and the main problem in cultivation is simply to keep it to a manageable size and going in the right direction.

ORIGIN
South America,
from Peru across to
southern Brazil

LONGEVITY
Long-lived scrambling
shrub and climber

SIZE
To 12 m (40 ft)

HABITAT
Woodland edge,
scrub, riverbanks

COLOURS
Species is purple;
cultivars may be red,
white or dull orange

A single bougainvillea flower is relatively inconspicuous compared to the showy bracts.

The pseudobulbs (storage organs) and young shoot of Cattleya.

Cattleya purpurata (Orchidaceae)

CATTLEYA ORCHID
Symbol of luxury

Their expansive flowers, with provocatively sexual lip and often strong colours, make cattleyas the quintessential orchid. In the late nineteenth century, when orchid-growing took off as an elite hobby, they were worn as buttonholes in expensive clothes, and associated with cigars, boxes of chocolates and vintage wines. If someone wanted to express luxury to advertise a product using an orchid, they would choose a cattleya.

Found growing as epiphytes (i.e. attached to tree branches or trunks) in hilly areas of Central and South America, cattleyas and the continent's many other orchid genera get all the water they need from direct rainfall, and their nutrients from leaves that fall on to the base of the plant. It is a lean life, often involving long periods of dormancy. Flowering tends to be spectacular and targeted towards being pollinated by very particular insect species. Our species is a native of southern Brazil, and is regarded as the national flower. However, its numbers and those of many other species have been greatly diminished not just by the destruction of their habitat but also by voracious collecting during the nineteenth century.

This orchid became a fashion item for the wealthy of western Europe and, to a lesser extent, the United States, and there was no holding back the greed of the collectors sent out to plunder the forests of South America. Trees were felled simply to strip the orchids off the branches. Collectors would sometimes destroy plants they could not carry to stop rivals getting hold of them. Bundled up in baskets and boxes, they would be transported to Europe, where only a small proportion survived more than a few months. During World War I many collections were lost to gardeners going off to fight and coal being used for purposes other than heating greenhouses. Once known as the 'Queen of Orchids', cattleyas have been displaced by *Phalaenopsis* (see page 129) and are now grown only by dedicated hobbyists. Unless, that is, you live in a climate where they can simply be hung up in wooden baskets and live off the rain.

ORIGIN
Southernmost
Brazilian states

LONGEVITY
Long-lived perennial

SIZE
50 cm (20 in),
when in flower

HABITAT
Upper branches of trees
in seasonally wet forests

COLOURS
This species: ivory and
purple; others may be
yellow, orange, pink

Delphinium 'Jubelruf' (Ranunculaceae)

DELPHINIUM

Cool blues

True blues, as opposed to purple- or violet-blues, are rare among flowers. Many delphiniums, however, are true blue, and on a very generous scale. Their size and the intensity and purity of their colour fascinate growers. Plants of cool-summer northern regions, they grow rapidly during long summer days, and indeed it is said that the very best place to grow them is just north of the Arctic Circle.

Delphiniums have come and gone in fashion. Our example is one bred by Karl Foerster, one of the most important figures in garden history in the early twentieth century. Interwar Germany had a fascination with the colour blue; it was seen by artists and designers as having mystical qualities, and delphiniums were a perfect expression of the colour for the garden.

Pioneering French botanist Carolus Clusius described *D. elatum* in 1601, but little serious cultivation happened until the late nineteenth century, when systematic breeders, such as Victor Lemoine of Nancy in eastern France, took an interest in the plants, creating crosses between different species and selecting progressively bigger plants with fuller flowers. The grand herbaceous borders beloved of British and German gardeners from the turn of the twentieth century onwards were ideal territory for delphiniums, high-maintenance plants that could be ministered to constantly by the armies of gardeners the low-wage economies of the time could support.

The hotter summers of the United States rendered many European cultivars short-lived. Frank Reinelt, a Czech immigrant to California who had been head gardener to Queen Marie of Romania, developed a series of seed strains in the 1930s, producing plants for a wide range of climatic conditions. He went on to try to breed pink and red varieties, since these colours are available in the generous gene pool of *Delphinium*. Although he was successful, these have never achieved the popularity of the blue cultivars, and interest in them today is largely from the floristry industry.

ORIGIN
Russia (core gene pool); species from North America added over time

LONGEVITY
Medium-lived perennial

SIZE
To 2 m (6 ft)

HABITAT
Open forests, woodland edges, damp hollows in steppe

COLOURS
Classically blue, but pinks, reds, yellows all exist

The seed case of a delphinium contains shiny jet-black seeds.

*Like the bougainvillea, the real flower of
a poinsettia is tiny and inconspicuous.*

Euphorbia pulcherrima (Euphorbiaceae)

POINSETTIA

Holiday offering

A modern Christmas would not be the same without the bold red flowers of poinsettias. As with all *Euphorbia*, however, the plant's showy 'petals' are in fact bracts, the flowers being the small yellow-and-green structures in the middle. The flower was held in high regard by pre-Columbian Mexican civilizations, and was used in traditional medicine, but was soon taken over by Christianity after the Spanish conquest, being included in nativity displays by Franciscan friars. A popular legend tells of a poor girl who could not afford flowers to take to the church; she took some roadside weeds instead, which turned into scarlet poinsettias at the altar.

The plant gets its name from Joel Roberts Poinsett, an American diplomat and botanist who first grew the plants in the United States in a greenhouse in South Carolina in the 1820s. Commercial production began in the early twentieth century with Albert Ecke, a German immigrant to California who saw the possibilities of the plant and researched how to grow it. His son developed a grafting technique for producing flowering plants at a more compact size (naturally these are rather leggy shrubs), which enabled them to be marketed successfully as pot plants; the family managed to keep the details of the technique a secret until the 1980s. There is an additional trick: tissue with a bacterial infection is necessary in the flowering shoots to weaken the main bud and stimulate flowering side-shoots. The Ecke family were very effective promoters of the plant from the early days, taking advantage, for example, of the development of television in the 1950s and 1960s to promote their plants by sending poinsettias to all the well-known presenters.

Poinsettia production is a major industry nowadays, and in fact the plant is the world's most economically significant potted plant. Breeding has resulted in a great many improved varieties, including some with cream or pink bracts, but the classic red is still overwhelmingly the most popular.

ORIGIN
Mexico

LONGEVITY
Long-lived shrub

SIZE
3 m (10 ft)

HABITAT
Open woodland

COLOUR
This species: scarlet; others cream or pink

Fuchsia magellanica (Onagraceae)

FUCHSIA

Nurseryman's dream

There are about 100 wild species of *Fuchsia*, from which a handful have been used to produce some 12,000 cultivars since the introduction of the first plants to Europe in the late eighteenth century. Ours, *F. magellanica*, is relatively hardy, but mystery surrounds its first appearance in cultivation. There is a story about the prominent 18th-century London nurseryman James Lee seeing a plant in the window of a rather humble house and having to pay the householder a substantial sum for it. However, there is also a suggestion that Lee concocted this story to cover up the reality of the plant having been stolen from the Royal Botanic Gardens at Kew.

The appearance of the less hardy *F. fulgens* in 1837 stimulated a veritable breeding frenzy. The plants are easy to grow and to propagate, and every seedling from a cross looks good, making it a nurseryman's dream: mass production *and* innovation. By 1848 a French book listed 520 varieties and 40 years later another listed 1,500, while thousands of plants were being sold every day in the markets of Paris and London during the summer. Because of their prominent pendant flower, fuchsias are well-suited to creating striking mass effects; for example, the cast-iron pillars of the Crystal Palace in London, built to house the Great Exhibition of 1851, were clothed in them for special events.

Today, fuchsias are still common, but only enthusiasts grow them in any quantity. Traditionally they have been a working-class flower, an opportunity to grow something exotic that does not need much extra heat to keep it going over the winter. Hobby growers have societies dedicated to their cultivation and organize competitive shows to enable their members to display their plants and their expertise in growing them. These events are often a part of one of the larger flower and produce shows traditionally held during the summer in Britain, always under huge tents set up on a field.

ORIGIN
This species: tip of South America; more generally, mostly South America, but also some in Mexico and the Caribbean, New Zealand and some Pacific Islands

LONGEVITY
Long-lived shrub

SIZE
3 m (10 ft)

HABITAT
Woodland, scrub

COLOURS
Dark pink-red with purple; cultivated forms mix red/pink and purple in endless combinations; no blues, yellows or pure whites

Fuchsia berries are small and
usually dark red, and can
be used to make jam.

*The bud emerging from the bulb
is full of decadent promise.*

Hippeastrum 'Royal Velvet' (Amaryllidaceae)

HIPPEASTRUM

Box of delight

Most people's experience of this plant starts with a hippeastrum – or amaryllis, as they are still often called – in a cube-shaped cardboard box. These are quintessential gift plants, performing the magic of starting as a satisfyingly big dry bulb and turning, as if by magic, with the minimum of care, into a spectacular head of flowers atop a sturdy stem, all without any encumbrance of foliage until after the flowers have died.

The name amaryllis is that of a genus of plants long since split up, and now restricted to a South African genus of similar-looking plants. Early botanists had few clearly agreed rules for how to classify plants, and were often overwhelmed by the vast amount of new material coming in from those great stretches of the Earth that Europeans were busy exploring and colonizing. Names had to be given, and fast. *Amaryllis* was a designation by Carl Linnaeus himself, the founder of the whole system of naming life, in the late eighteenth century. One of those responsible for sorting out the mess with *Amaryllis* was William Herbert, whom we met earlier as a pioneering breeder of daffodils (see page 50). However, disagreements over *Amaryllis* and *Hippeastrum* continued until the 14th International Botanical Congress in 1987, which finally settled the matter in favour of the latter.

Hippeastrums, often fondly dubbed 'hippies', are not just one-shot wonders. If potted up and put outside for the summer they will grow a bunch of strap-shaped leaves and flower again, at least if brought back into the warm. The plants grown today as houseplants are all hybrids, bred from a few of the 90 species found in the wild. The first hybrid was raised in 1799 by a hobby gardener, Arthur Johnson, a watchmaker from northern England. His plants ended up in the Liverpool Botanic Gardens, and then were taken up by nurseries elsewhere, including, crucially, by the innovative de Graaff family of nurserymen in mid-nineteenth-century Holland. That country's greenhouses are still a major producer of the plant.

ORIGIN
Southern South America

LONGEVITY
Long-lived bulb

SIZE
Varies; about 60 cm
(24 in) when in flower

HABITAT
Woodland, scrub; some species grow as epiphytes (i.e. grows on another plant) or on rocks

COLOURS
Dark red to white

Lathyrus odoratus 'Cupani' (Fabaceae)

SWEET PEA

Scent of summer

Familiar as a delicate, short-lived cut flower, and therefore never to be seen in those flower shops that sell only what is mass-produced in glasshouses and trucked a thousand kilometres to its point of sale, the sweet pea is a real enthusiasts' flower. It is easy to grow but requires a certain amount of organization and dedication, the result of which is the enrichment of several months of early summer by a succession of exquisitely sculpted flowers in a variety of colours, often with a lingering sweet fragrance. Indeed, for many, the fragrance is one of the main reasons for growing it.

This particular one was introduced to Britain by a Sicilian monk, Friar Franciscus Cupani, who in 1699 sent seeds to a correspondent there, a teacher from Enfield, north of London. This became the ancestor of today's sweet pea, but not for about a century. It was only in the late Victorian period, when a number of amateur growers and head gardeners began to make crosses with other species, that the modern sweet-pea era began. In 1900 some 264 varieties were shown at the Bi-centenary Sweet Pea Exhibition, held in London's Crystal Palace. The National Sweet Pea Society was founded the next year. One particular company, Unwins, which started out selling flowers to London markets, specialized in sweet peas and made many breakthroughs in colour and form, and the firm is still important today.

Gardeners and nurseries on the east coast of the United States started breeding sweet peas almost as soon as they arrived there, in the late nineteenth century, while Californian breeders began working with the plant in the 1910s. Central Valley once had vast fields for seed production, but now that gardeners buy starter packs of young plants rather than seed, this has been greatly reduced. A leading sweet-pea breeder for much of the twentieth century was Anton Zvolanek in New Jersey, who introduced multi-stemmed and semi-dwarf varieties. Contemporary breeding has focused on resistance to disease, and on dwarf tendril-less varieties for small-scale use or containers.

ORIGIN
Southern Europe (key species); others in the gene pool are from other cool temperate climates, in Asia and the Americas

LONGEVITY
Annual

SIZE
Rarely more than 2 m (6 ft)

HABITAT
Woodland edge, scrub

COLOURS
Pinks, violet-blues, creams

*The distinctly hairy immature
seed pod of a sweet pea.*

As with many members of the daisy
family, the seeds of edelweiss have
fine hairs to catch the wind.

Leontopodium nivale (Asteraceae)

EDELWEISS

Blossom of snow

This is one of those flowers that is genuinely 'iconic', used so extensively to promote the European Alps that the mere sight or mention of it will say 'Switzerland' or 'Austria' to most people. Along with the blue of gentians (see page 106), this is the flower that adorns everything from chocolate boxes to men's shirts.

The flowers of *Leontopodium* (there are a handful of species) are certainly distinctive, with a dense coating of fine white hairs, probably a defence against desiccation, predation and the ultraviolet light that is particularly strong and damaging at high altitude. But in fact, as with bougainvillea, poinsettia and quite a few other flowers that we feature in this book, the 'flowers' are really flower clusters, the furry 'petals' being bracts to attract pollinators. The fact that they are bracts and densely hairy does mean that they are relatively durable, and hence long-lasting enough to be included in souvenirs of the alpine regions.

It was during the late nineteenth century that edelweiss acquired its symbolic values, a period when the various European ethnic groups were searching for symbols and traditions in order to define themselves. Edelweiss (which means 'noble white' in German) came to symbolize the purity of the alpine regions. The flower became an emblem of the Austro-Hungarian Imperial couple Franz-Josef and Elisabeth, the latter sometimes wearing jewelled versions in her hair. It also became the symbol of a number of regiments in the Imperial army, featuring on various military uniforms, and – with the rise in interest in hiking and mountaineering – of a number of outdoor sports clubs. Not surprisingly, many of the flowers were picked, causing the populations to decline. Strict laws protecting wild plant species became a feature of most alpine regions in the twentieth century, and during these decades edelweiss iconography boomed. It was used both by units of the Nazi military and by an anti-fascist youth group, the Edelweiss Pirates; it has been co-opted by airlines and tourist agencies, has appeared on banknotes and has featured in innumerable songs and films.

ORIGIN
Mountains of Eurasia

LONGEVITY
Short-lived perennial

SIZE
20 cm (8 in)

HABITAT
Rocky meadows, scree slopes, usually on limestone

COLOUR
Off-white

Pelargonium 'Balcon Red' (Geraniaceae)

GERANIUM, PELARGONIUM

Mediterranean balconies

The so-called geraniums, long reclassified as *Pelargonium*, are a supremely successful group of plants – if success is determined by how widely they are grown. Our example is derived from *P. peltatum*, an almost succulent drought-tolerant plant that naturally scrambles through shrubs but in cultivation hangs happily downwards, festooning countless balconies and window boxes across German-speaking central Europe.

It was, however, the bushier 'Zonal' pelargoniums, also with dramatic scarlet flowers, that established the plants' reputation in the nineteenth century. They were on the one hand ideal for mass displays in public parks and, on the other, tough enough to survive on more or less anyone's windowsill. Simple red zonal pelargoniums became the poor person's houseplant par excellence; indeed, for many people a pelargonium was a status symbol, something living that did not require a garden and could be nurtured, almost like a pet. It is very likely that few working-class growers ever bought their plants, since they are easy from cuttings, so family and friendship networks would have distributed them widely.

Pelargoniums had been introduced to Europe by the Dutch in the late seventeenth century, and by the early nineteenth there were a great many hybrids in the greenhouses of the wealthy. By the end of the century the simple red 'geranium' was everywhere, including Switzerland. Official encouragement as a recession mood-breaker came in 1936, when the city council of the capital, Bern, started a campaign to festoon the city with them. By the late 1940s the trailing 'Balcon' types had taken over, being better suited than the upright zonals to life on the window ledge. In 2016 several institutions collaborated to declare Bern 'Geranium City', with special displays. However, the spread of the plant into window boxes on the typical wooden chalet houses of the Swiss countryside had peaked by this time, with rival plants appearing, especially new strains of trailing petunia. The windows of central Europe may become a horticultural battleground.

ORIGIN
South Africa

LONGEVITY
Long-lived woody
perennial

SIZE
Parent species can climb
or trail to 2 m (6 ft)

HABITAT
Scrub

COLOURS
Scarlet; white and
pink garden forms

The seed capsule has a spring
mechanism that discharges the seed;
it is seen here after the
seed has been ejected.

*A petunia seed pod breaks open
to release the tiny seeds.*

Petunia 'Swiss Dancer' (Solanaceae)

PETUNIA

Worth investing in

Petunias have become one of the most familiar plants for summer displays. Familiar, but also rapidly changing, as new varieties seem to come out with remarkable frequency. Free-flowering and easy to grow, they have attracted the kind of financial investment into the science of their genetics that normally only edible crops would garner. The roots of this may lie in their having played an important role during the early days of research into viruses, since they are prone to the tobacco mosaic virus, which was the first to be scientifically understood.

The familiar petunia was initially a hybrid between two species from Uruguay, introduced to Europe in the early nineteenth century. It spread very quickly as a summer bedding plant, and breeders in England, Germany, France and Belgium developed many varieties in the 1840s and 1850s. In the early days petunias were propagated by cuttings, but with the understanding of Mendelian genetics after 1900 seed production became possible. German breeders led the way, and in the 1930s they produced the Grandiflora varieties, with big but easily damaged flowers, and the Multifloras, more robust and with smaller flowers in larger numbers.

Japanese breeders became involved early in the twentieth century, and are still. Intriguingly, it is two distilleries – Kirin and Suntory – that have become major players. Breeders in the United States have been significant, too, and the PanAmerican Seed company produced many innovations, including the first true red, 'Comanche', in 1953.

Recent decades have brought a concentration on trailing varieties, which are better for ground-cover, hanging baskets and window boxes. Adventurous crosses have been made, and species from related genera, notably *Calibrachoa*, have been pulled into the gene pool through laboratory-led breeding. New varieties tend to be grouped in 'series', with breeders trying to produce as many colours as possible in a particular format. Examples are the Cascadias, Conchita, Tumbelina, Fanfare, Designer, Potunia and Madness series. Among the most successful have been the trailing Surfina varieties developed by Suntory.

ORIGIN
Southern South America

LONGEVITY
Short-lived perennial, usually grown as annual

SIZE
30 cm (12 in)

HABITAT
Open country: grassland, savannah, woodland edge, coastal areas

COLOURS
Pinks, reds, purples, whites; no true blues, not many yellows

Protea cyanoides (Proteaceae)

KING PROTEA

Phoenix flower

Huge flower heads (sometimes up to 30 cm/12 in across) make this the most impressive of an impressive genus. Small flowers are packed into heads surrounded by colourful bracts that attract the plants' pollinators, in this case sunbirds, which drink the copious nectar the flowers produce. This species is also the most widely distributed protea. Not surprisingly, it has become the South African national flower, its most familiar symbol after the outline of Cape Province's Table Mountain, and appears as an emblem on countless logos, insignia and badges.

Protea and the protea family are ancient, dating back 140 million years to before the split-up of the southern continent of Gondwana. Many protea species may be much younger, however, a result of the extraordinary burst of evolution that has occurred in South Africa, and in Cape Province in particular. It is now thought that much of this evolution was stimulated by the need for plants to adapt to periodic fires. This particular species has a lignotuber, a large underground organ that survives fires and resprouts in the winter rains. Its seeds also germinate after fire, along with those of a great many other plants of the extraordinarily rich *fynbos* habitat of the region.

Proteas first appeared in European greenhouses in the early nineteenth century, but they did not survive, since they have somewhat unconventional cultivation needs that were not understood until much later. The Cape heaths, closely related to the heathers of European moorlands, did much better, and for a brief period were extremely popular. The king protea's day came later, in the latter part of the twentieth century, when it and various other members of the family became important plants for the global floristry industry. The flowers are not just impressive, but long-lasting and physically robust, and ideal for a certain type of grand bouquet. They are now grown commercially, not just in South Africa, but in many other dry-summer climate zones, especially on poor acidic soils that resemble those of the places where they grow naturally.

ORIGIN
Cape Province,
South Africa

LONGEVITY
Long-lived shrub

SIZE
2 m (6 ft)

HABITAT
Fynbos, a dense
vegetation similar to
Mediterranean *maquis*,
much of it no higher
than head height

COLOURS
Pink, pale yellow

Protea seeds are distributed on the wind, often to germinate after a bush fire.

Flower buds form in summer,
overwinter, and then burst
open the following spring.

Rhododendron 'Cynthia' (Ericaceae)

RHODODENDRON

Victorian choice

For just a few months at the border of spring and summer, rhododendrons can dominate gardens and landscapes – or be completely absent! These are plants with particular requirements: not too cold or too hot, plentiful rainfall and acidic soil. Where they are happy they are very happy, and when they are not, they simply won't grow. There are more than a thousand species, ranging from tree-size down to tough, low carpeting plants, but in cultivation it is the big shrubby ones that are most common.

Rhododendrons became something of an obsession for elite gardeners at the turn of the twentieth century. This was the period when British (and a few American) plant-hunters were exploring the Tibet/China/Burma border region, where this particular genus has had one of those explosions of evolution that nature occasionally exhibits; there are hundreds of species in this area. These plants were introduced as seed at a time when a confident British upper-middle class was moving out of London into the countryside, particularly on to poor, acidic soils where the classic English garden plants, such as roses, did not flourish. Rhododendrons, however, loved them.

Rhododendrons hybridize easily, and the first introductions, earlier in the nineteenth century, had been seized upon eagerly for breeding. Thousands of crosses were made, of which our 'Cynthia' was one of the most successful, particularly for the larger garden. Vast century-old plants now frequently dominate their surroundings, especially when smothered in hundreds of flower heads.

The downside of the plants is the dark evergreen foliage, and the absence of flowers after June. Many gardens dominated by the plants are dreary indeed for much of the summer. Rhododendrons are consequently unpopular with many, a situation that has been exacerbated by the invasive behaviour of one species, *R. ponticum*, which was often used as a rootstock for the hybrids, and consequently widely distributed. This species was once (before the last Ice Age) found across Europe, so its spreading behaviour is perhaps only to be expected.

ORIGIN
Much of North America, Eurasia and South East Asia; the ancestors of the garden hybrids are mostly of Himalayan origin

LONGEVITY
Long-lived shrubs

SIZE
To 10 m (33 ft) (this variety); others anything between 15 cm (6 in) and 20 m (66 ft)

HABITAT
Primarily woodland, in the shade of canopy trees

COLOURS
Pinks, reds, creams; some yellows, some purple-blues

Allium 'Globemaster' (Amaryllidaceae)

ALLIUM

Perfect spheres

As perfect a sphere as it is possible to have in nature, atop tall, upright stems: the visual drama of these ornamental garlics is terrific, especially when they are planted en masse. It is not surprising that, after many years of being something of a minority interest, these plants are attracting more and more attention from both designers and breeders. The latter are aiming for taller stems, bigger heads, richer colours.

Allium, the garlics, is a big genus – some botanists would say there are nearly a thousand species. All have the distinctive smell that is fundamental to so many cuisines, and all are potentially edible. While many are small, with loose flower heads, many have something approaching the big spherical heads that make these – often called 'drumstick garlics' – so distinctive. Let an onion 'go to seed', as vegetable gardeners would say, or even better a leek, and you have a similar dramatic flower head.

The drumstick garlics are overwhelmingly from the Middle East and Central Asia, climates with a short spring between the snows of winter and the baking heat of summer – the same region and climate as the far more familiar tulip. Unlike tulips and a number of other bulbs of the region, they did not attract the interest of Ottoman gardeners, and so their introduction to Europe was much later. For much of the twentieth century only specialist bulb catalogues listed them, in the 'Miscellaneous' section after the listings for daffodils, tulips and crocuses. Dutch breeders then began to work on them, producing varieties such as this 'Globemaster', with a big head and strong colour. New introductions are available only in small quantities at first, so prices are high. With time the plants 'bulk up', and the number available can increase almost exponentially. By the 1990s prices had come down, and with the timing of London's Chelsea Flower Show, held in May, being perfect for them, they were soon ultra-fashionable.

ORIGIN
Higher-altitude Central Asia: Iran across to Xinjiang, China

LONGEVITY
Short-lived bulb

SIZE
To just over 1 m (3 ft)

HABITAT
Open country, scrub, steppe

COLOUR
Purple .

An individual flower from an allium inflorescence (the technical term for a flower head).

One of the flowers from a flower head. Only through close observation can we make out the details.

Astilbe x *arendsii* 'Bressingham Beauty' (Saxifragaceae)

ASTILBE

Drawn to the damp

Fluffy but neat upright spikes in every shade from deep red to pure white adorn these plants, which are most often seen in the wetter spots of public parks and large gardens. Most of them really appreciate the damp, a fact that discourages many people from growing them; they are also a little on the slow side, but that is only because they are getting their roots in, and once established they will live, and slowly expand their clumps, for years. Astilbes flower only in early summer, but their foliage is a cut above that of other perennials and can be a real pleasure for the rest of the year. Their heyday was possibly the very beginning of the nineteenth century, when they were sold widely as temporary pot plants, very often 'forced', i.e. grown in greenhouses in late winter to produce early growth and flowers.

Our example commemorates an important man and an important place in twentieth-century garden history. Georg Arends, who worked in the German industrial city of Wuppertal in the early twentieth century, was a nurseryman who made a great many selections from seed-raised plants, or created good hybrids that have stood the test of time. He made crosses between a number of species and the hybrids that had been recently made by Victor Lemoine in France. The new plants he raised were gathered under the name 'x *arendsii*'. This use of hybrid names has been discontinued, however, and we now use just the cultivar name, which in this case refers to Bressingham Gardens, a nursery founded in Suffolk, eastern England, by Alan Bloom, who did perhaps more than anyone to lead a revival of perennial plants in Britain from the 1950s onwards. Bressingham was set up to show off the range of plants he was breeding and growing, as well as his nationally important collection of steam engines.

ORIGIN
Cool mountain areas in North America and eastern Asia

LONGEVITY
Long-lived perennial

SIZE
To 1 m (3 ft)

HABITAT
Stream sides, wet woodland, ravines

COLOURS
Purples, reds, pinks, white

Calluna vulgaris (Ericaceae)

HEATHER

Painting the hills purple

Associated strongly with Scotland – since it flourishes on the poor, acidic soil of this country's deforested hills, turning them spectacularly purple in midsummer – heather in fact thrives in many similar habitats in northern Europe. In Denmark, the Netherlands and Germany, however, many of the heathlands it used to dominate have been ploughed and fertilized to make way for something more useful. A plant of wild and barren places, the kind of territory our ancestors would usually have looked at with fear and trepidation, heather was never valued other than as a source of nectar for one of Europe's best honeys, until the very end of the nineteenth century.

Some innovative professional gardeners used heather for its evergreen nature to create temporary winter displays in the places they cared for, but it is probably fair to say that the polemical book *The Wild Garden*, published in 1871 by William Robinson, something of an ambitious journalist and self-made man, helped to launch its garden career and that of a number of other wild plants. He extolled their beauty and their virtues, which at this time were increasingly realized as a growing number of gardeners in the new suburbs around London were finding themselves on less than perfect soil. New nurseries sprang up to grow it and a number of other heather species, exploiting chance finds of plants with flowers of a different colour, or with foliage in something other than dark green. Heather gardens became popular, exploiting the fact that different species flower at different times; they were also very low-maintenance, which commended them to many.

There was a revival of interest in heathers in the 1960s in Britain, and the Heather Society of Great Britain was founded around this time. There followed a boom in specialist heather nurseries, with the English nurseryman Adrian Bloom, son of Alan (see previous page), bringing them to the peak of their popularity, combining them with conifers in island beds and promoting them for low-maintenance planting.

ORIGIN
Europe and Turkey

LONGEVITY
Long-lived subshrub

SIZE
50 cm (20 in)

HABITAT
Hillsides, open woodland, rocky places, always on acidic soil

COLOURS
Purples, pinks, white

One of the many tiny flowers that
make up each heather flower head.

The small seeds are attached to
fine hairs, known as thistledown,
to ensure long-distance travel.

Cirsium heterophyllum (Asteraceae)

MELANCHOLY THISTLE

Flower of Scotland

The thistle is the national flower of Scotland. The use of a plant with such a reputation for being thorny and weedy is an interesting one, and one on which the English (waving their rose) might pour scorn. It was the national flower by the fifteenth century and appeared in heraldry from then on, with a possible origin in a story dating back to the thirteenth century, when a group of marauding Vikings were creeping up barefoot on some Scottish soldiers – it is not difficult to imagine what happened next. The country's highest order of chivalry is the Most Ancient and Most Noble Order of the Thistle, founded by King James VII in 1687, with the Latin motto *Nemo me impune lacessit*, translating as 'No one can harm me unpunished'!

Those wanting botanical certainty will find no indication of what kind of thistle it was that originated this potent national symbol, but it might be interesting to consider one of the seven species of true thistle native to Scotland, *C. heterophyllum*, known as the melancholy thistle because of the way the flowers hang down from their stems. This is the only species that is not very spiny, and, given that it is one of the most attractive Scottish native plants, it is not surprising that this is the only one in garden cultivation.

The use of thistles as garden plants might strike some people as surprising, but there are a number that are very attractive and are being used increasingly in naturalistic or ecological planting styles. A handful of thistles from central Europe in particular are also being seen more and more in gardens. Our melancholy thistle is actually one of the best for this, since it spreads in an attractively loose way, not displacing other plants but steadily infiltrating the border. It is also a fine component of a wild-flower meadow planting, often flowering that little bit later than most meadow species.

ORIGIN
Northern Europe, Siberia

LONGEVITY
Long-lived perennial

SIZE
1 m (3 ft)

HABITAT
Moist grassland,
often on limestone

COLOUR
Purple-pink

Geranium x *oxonianum* 'Claridge Druce' (Geraniaceae)

HARDY GERANIUM

Covering the ground

Most of our popular garden perennials entered cultivation well over a century ago. The hardy geraniums, traditionally known as 'cranesbills', however, were largely ignored by gardeners until the last few decades of the twentieth century, since when they have become very popular indeed, especially among gardeners in cool temperate climates.

Our example is one of the very first hybrids to have been recognized, and is named after the amateur botanist who discovered it. George Claridge Druce was a pharmacist in Oxford who was famed for his hangover cures (surely much needed in a city of hard-drinking dons and students). The plant, a hybrid between two southern European species, first occurred in Dr Druce's garden in 1900. It was not, however, formally given its name until the early 1960s, thanks to Graham Stuart Thomas, the garden supremo of Britain's leading conservation charity, the National Trust. It is a very vigorous plant, good at suppressing weeds and seeding itself around, but never becoming invasive. As do its parents, *G. endressii* and *G. versicolor*, it has the remarkable quality of growing at very low temperatures, which makes it ideal for the climate of the British Isles, because it grows through the winter while most other perennials are dormant. It is just the kind of plant that Thomas was on the lookout for, something that could cover a lot of ground decoratively, since his main problem was how to conserve all the vast acreages of historic gardens using a workforce that was a fraction of what it would have been historically.

Thomas was certainly one of those whose practice and eloquent writing helped to promote geraniums generally. A huge revival of interest in perennials in Britain for a time had the vigorous, easy and colourful *Geranium* as its focus, and there were a great many new introductions and new varieties developed; gardeners in the United States and the Netherlands soon followed suit.

ORIGIN
Southern Europe
(both parents)

LONGEVITY
Long-lived perennial

SIZE
80 cm (32 in)

HABITAT
Parent species:
woodland edge

COLOUR
Pink

The seed head, with five seeds and springs ready to fire, which they do as they dry, releasing the seeds.

*Individual disc florets like these
make up the centre of the
gerbera flower head.*

Gerbera 'Sweet Glow' (Asteraceae)

GERBERA

Mother's herb

The sea of large daisies atop their tall, stiff stems stretch off in all directions, occasionally changing colour as another variety takes over – and all under glass. Welcome to one of the growers that makes up the nearly 200 hectares (495 acres) of gerbera under glass in the Netherlands, the world's biggest producer of this most modern of flowers. Gerberas are currently a florist's favourite, long-stemmed, long-lasting, easy to transport, colourful and immensely varied (so long as you don't want blue). Along with the tulip, they are among the most 'artificial' of flowers, with their simple forms available in a seemingly endless number of variations, ideal for minimalist contemporary interiors.

Traditionally used as a herb by nursing mothers in South Africa, gerberas have made an extraordinary and relatively recent leap into modernity. They are easy to hybridize and, unlike with many plants, a large proportion of the seedlings from a cross will look not just good, but also distinctive. Put these characteristics together with modern genetics and laboratory techniques, and you have a breeder's dream, a plant that is extraordinarily malleable. The range of colours, colour combinations and petal shapes is incredible, and this, when cross-referenced against a range of flower sizes, results in thousands of possibilities. They are also relatively easy to grow and perfect for mountain regions in the tropics, where equable temperatures combine with good light to create ideal conditions for production; many are now grown in Colombia, Ecuador and India, mostly for export.

However, like all crops, gerberas are prone to pests and diseases, the worst being powdery mildew, which can devastate crops rapidly. The treatment requires pesticides, and since the plants need intensive handling (as do all floristry crops), the exposure of the workforce to chemicals has created a big health problem. The answer probably lies with crop genetics, identifying the genes in those plants that are resistant to the mildew, and then bringing those genes into the decorative varieties the market wants, so that the use of chemicals can be minimized. The future looks bright indeed for gerberas.

ORIGIN
South Africa

LONGEVITY
Long-lived perennial

SIZE
60 cm (24 in)

HABITAT
Grassland, open woodland, scrub

COLOURS
Naturally orange to yellow and white; cultivars can be almost black, along with pink and many 'hot' shades

Index

Note: common names are printed in plain type; species names are printed in italics.

The Author

NOEL KINGSBURY is internationally known as an innovator, writer
and teacher in the fields of gardening and planting design, especially
in researching and promoting naturalistic planting. He has also written
about the history of agricultural crops and garden plants.
Current projects include a global education portal for garden people
and an experimental low-irrigation garden in Portugal.

Acknowledgements

Much of this represents the outcome of a lifetime learning about flowers
and their fascinating histories, so I owe a debt of thanks to the various
unnamed librarians, researchers and writers who have helped along
the way. Amongst writer-researchers I have drawn heavily on, in
particular Alice M. Coats (1905–78), whose erudite books written
in the 1950s are still important sources and Heinz-Dieter Krausch
(1928–2020) whose *Kaiserkron und Päonien rot*, is simply the best single
source there is. I owe thanks to Malgorzata Kiedrzynska who helped me
with some research, and to the support of my wife, Jo Eliot, who gets
a running commentary on all that I work on.